Periodic Poetry

an unstable collection

by Mz. Donkey

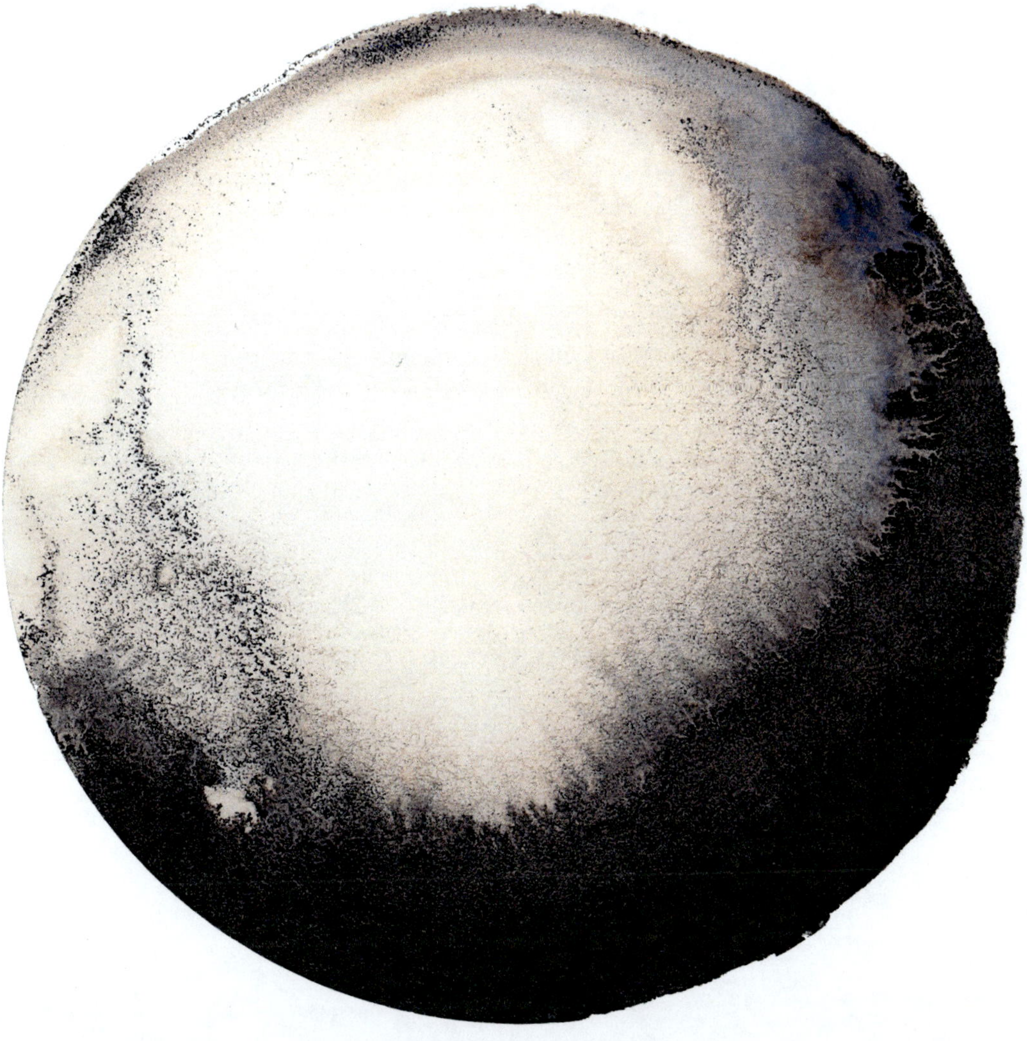

Periodic Poetry: An Unstable Collection

Published by Donkey Drove

This book is a work of creative nonfiction. Scientific facts are accurate to the best of the author's knowledge. For bonus weirdness, visit mzdonkey.com/unstablebonus

First Edition

For my little boos. Stay weird.

Table Of Contents

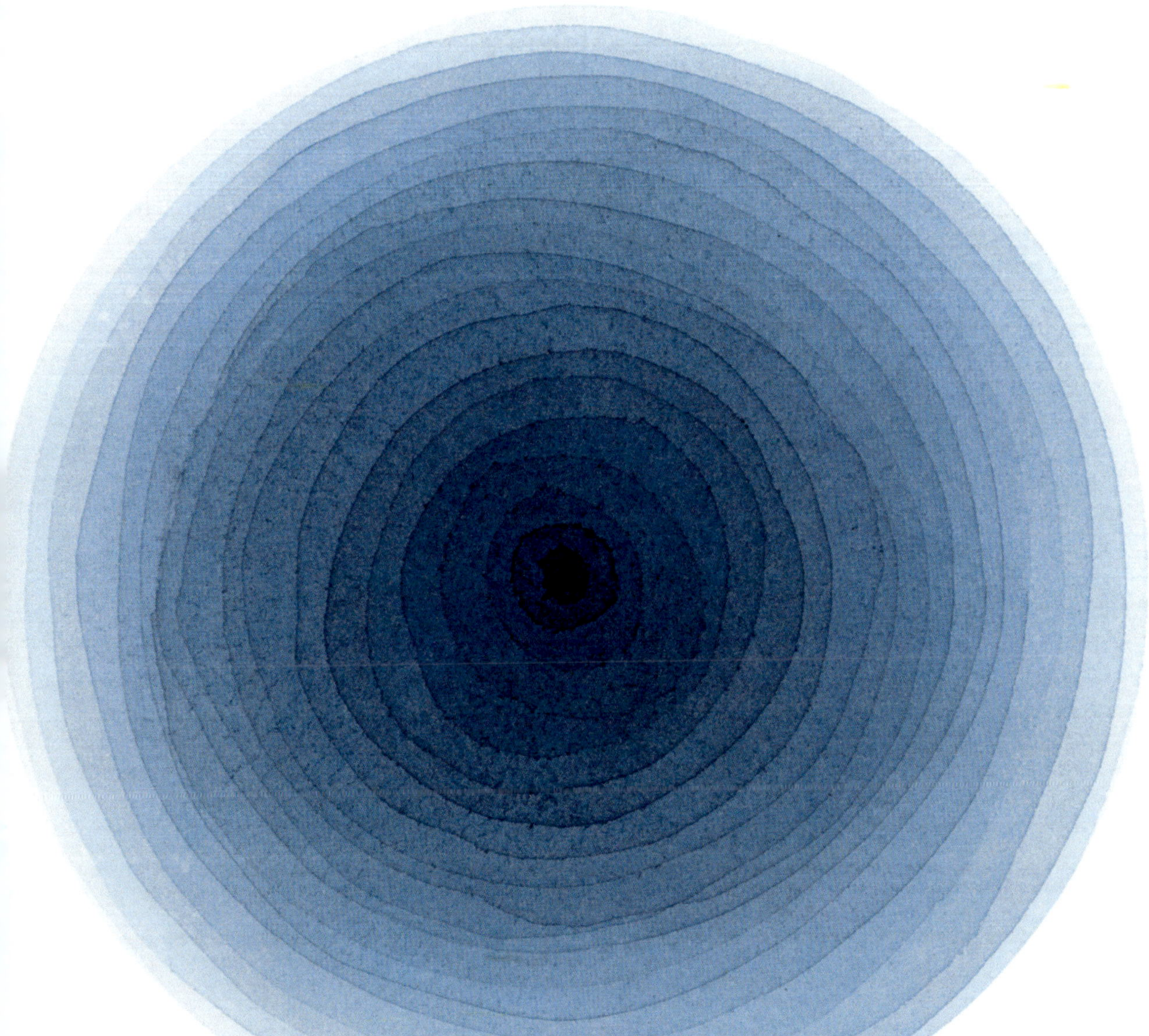

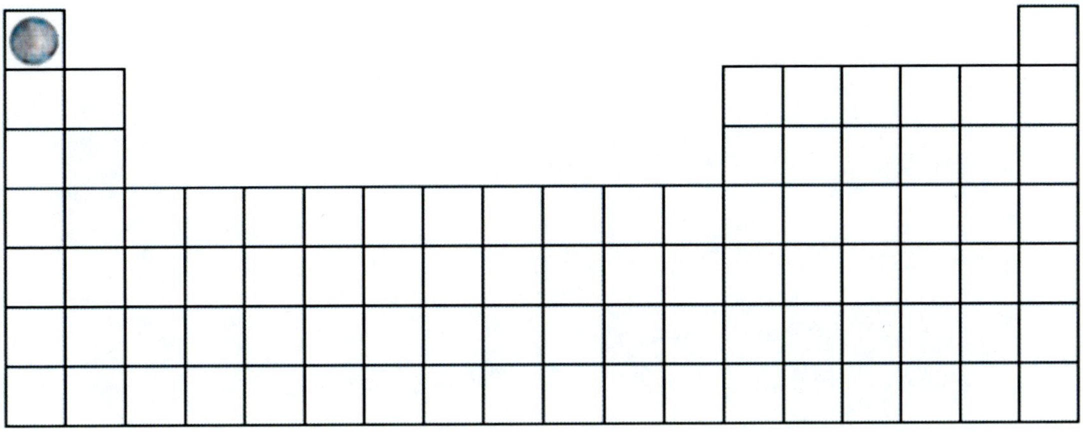

Hydrogen

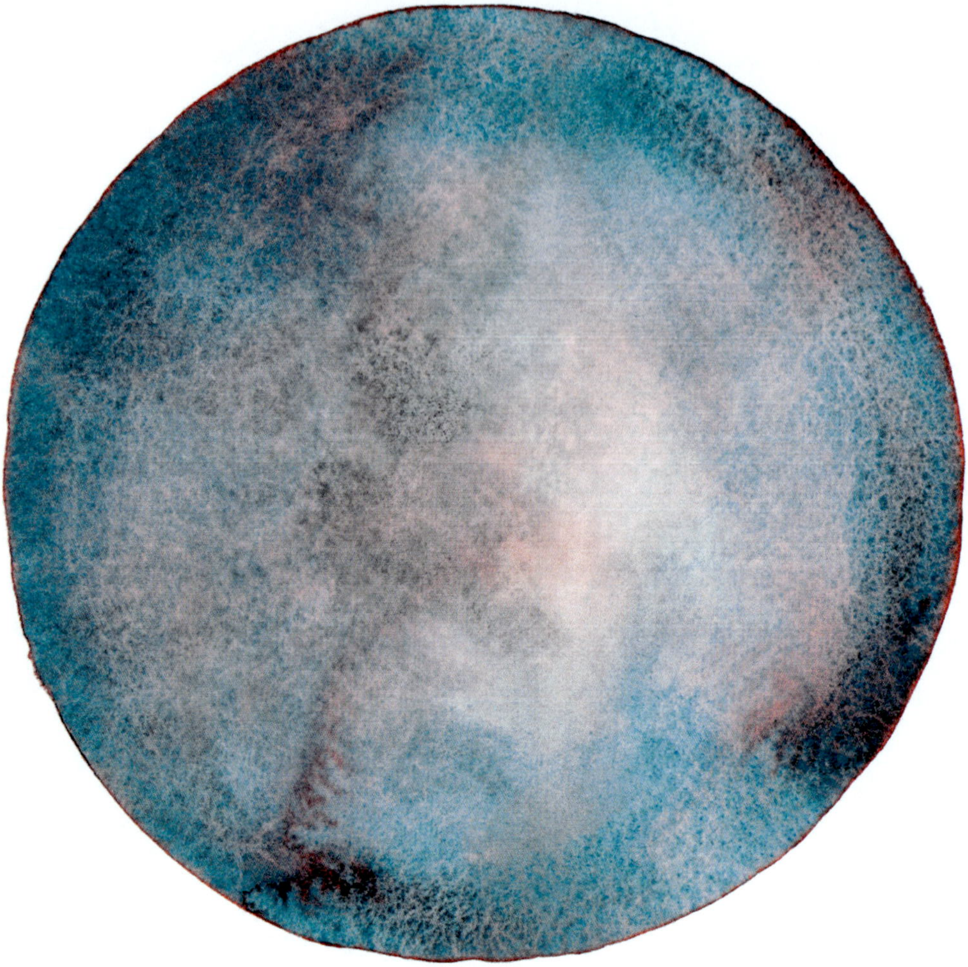

H

Hydrogen's in a class all its own

#1: Friendly, abundant, and light

Super flammable (Peace, Hindenburg)

Tons of energy, vital for life

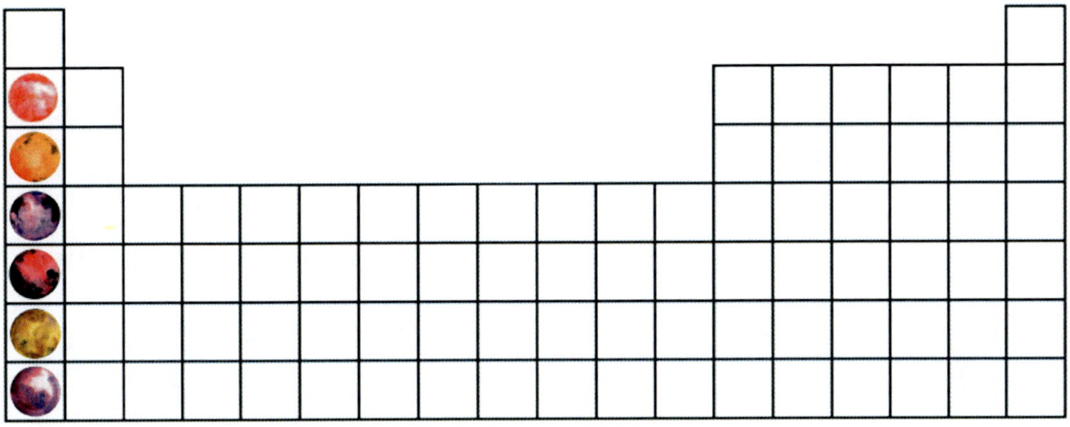

Group One

Alkali Metals, one outer electron

Burn to reveal their identity

Highly explosive (best stored in oil!)

They're soft and have a low density

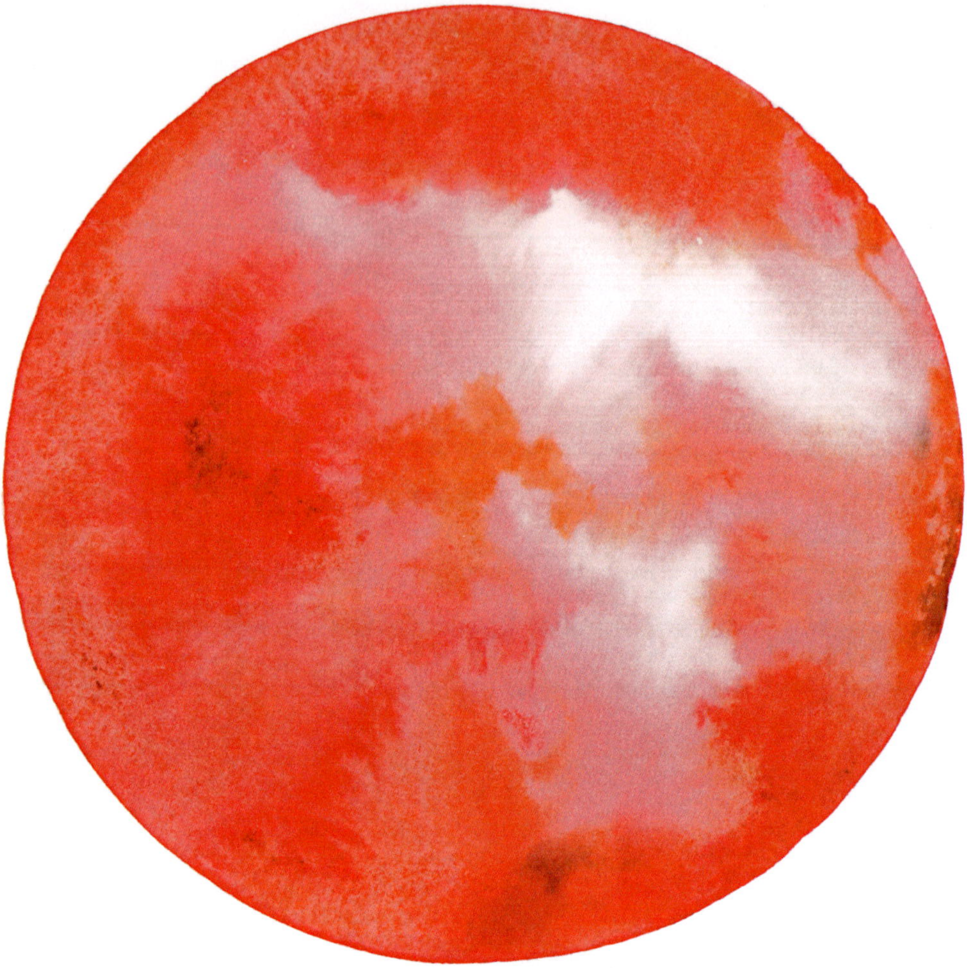

Li

Lithium is number 3

So light it dances on water

Makes quality grease and batteries

Helps smooth out a mental disorder

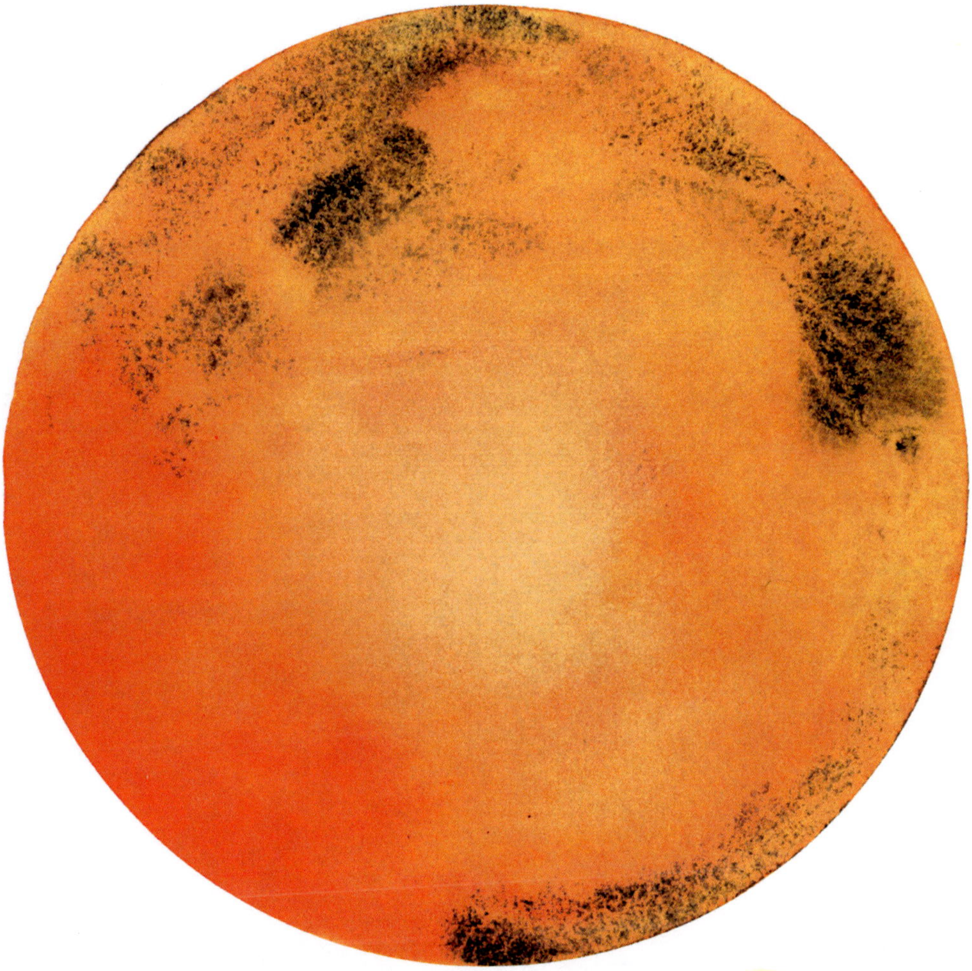

Na

Number 11 gives us sodium

It’s the reason the oceans are salty

Creates fire when exploding in water

Crucial to balance biology

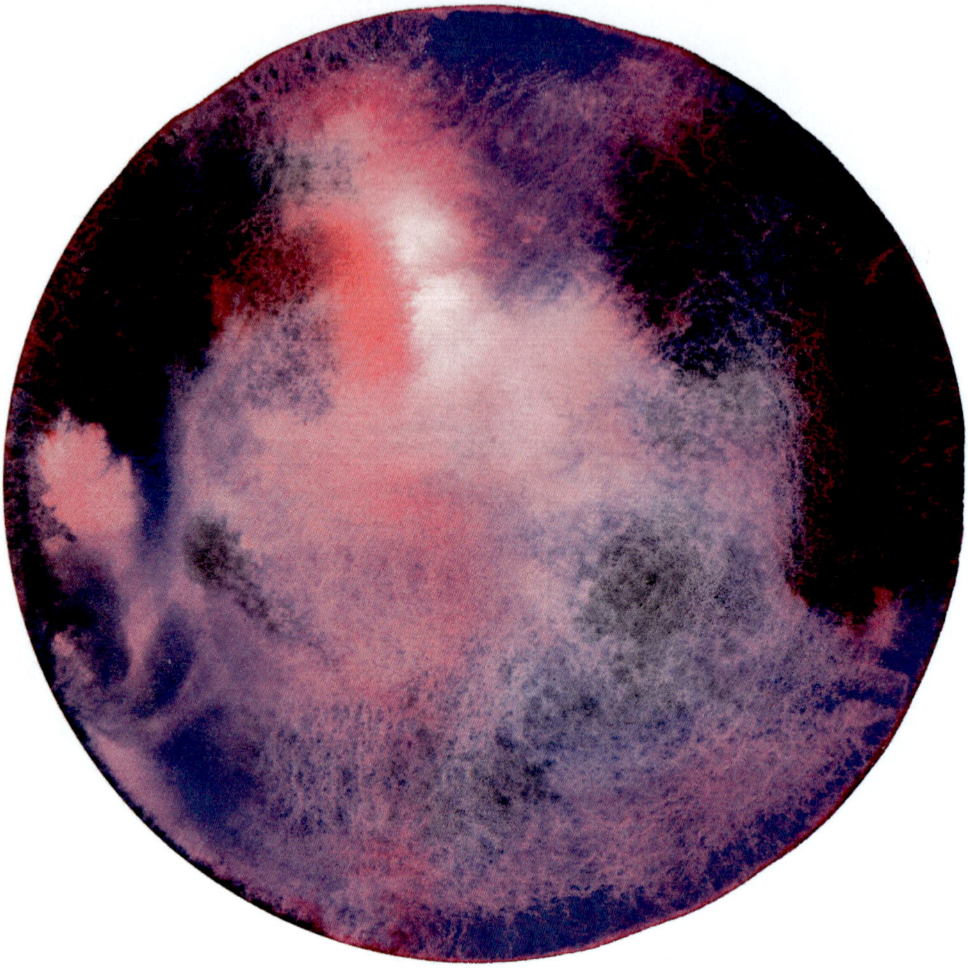

K

Number 19 is potassium

Essential for heartbeats and nerves

Ingest too much - it'll kill you

It's lavender when it burns

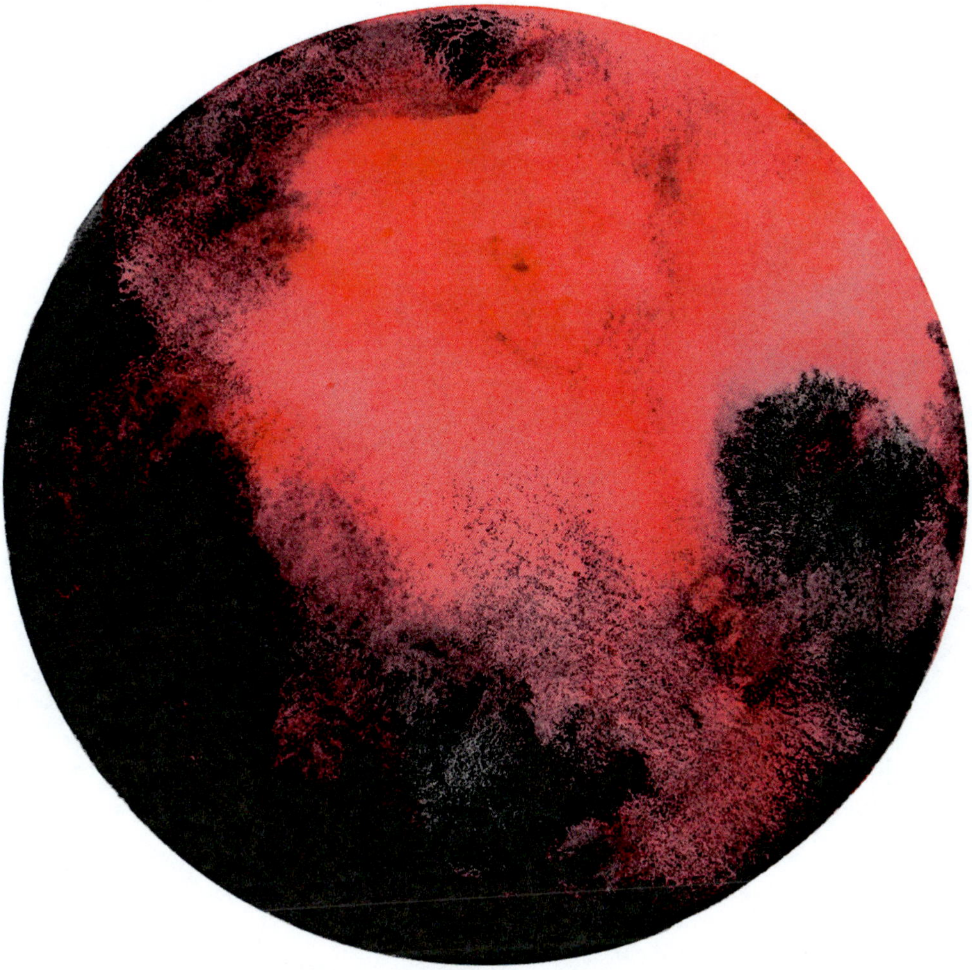

Rb

37 Rubidium

It's a propellant in space

Spontaneous combustion and lasers

Burns with a red-violet flame

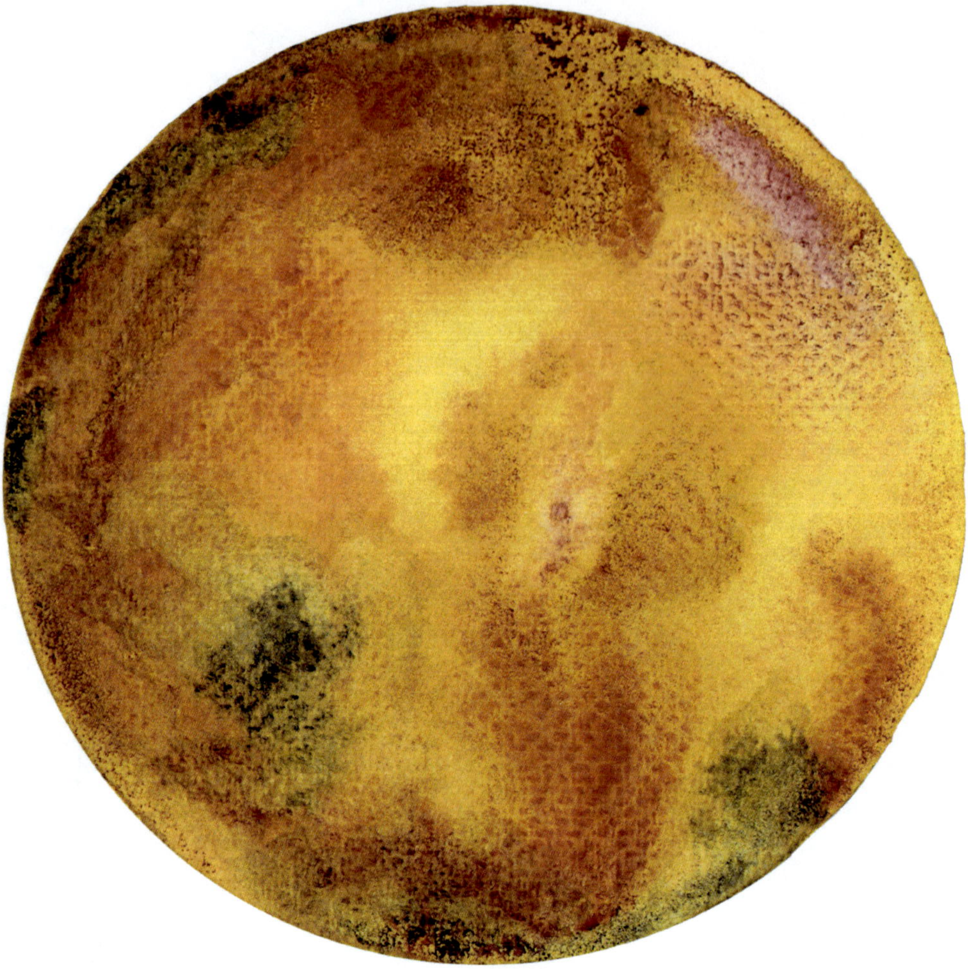

Cs

Element 55 is cesium
It's the atomic clock gold standard
Melts in your hand, *extreme* reactivity
Night goggles and helps you fight cancer

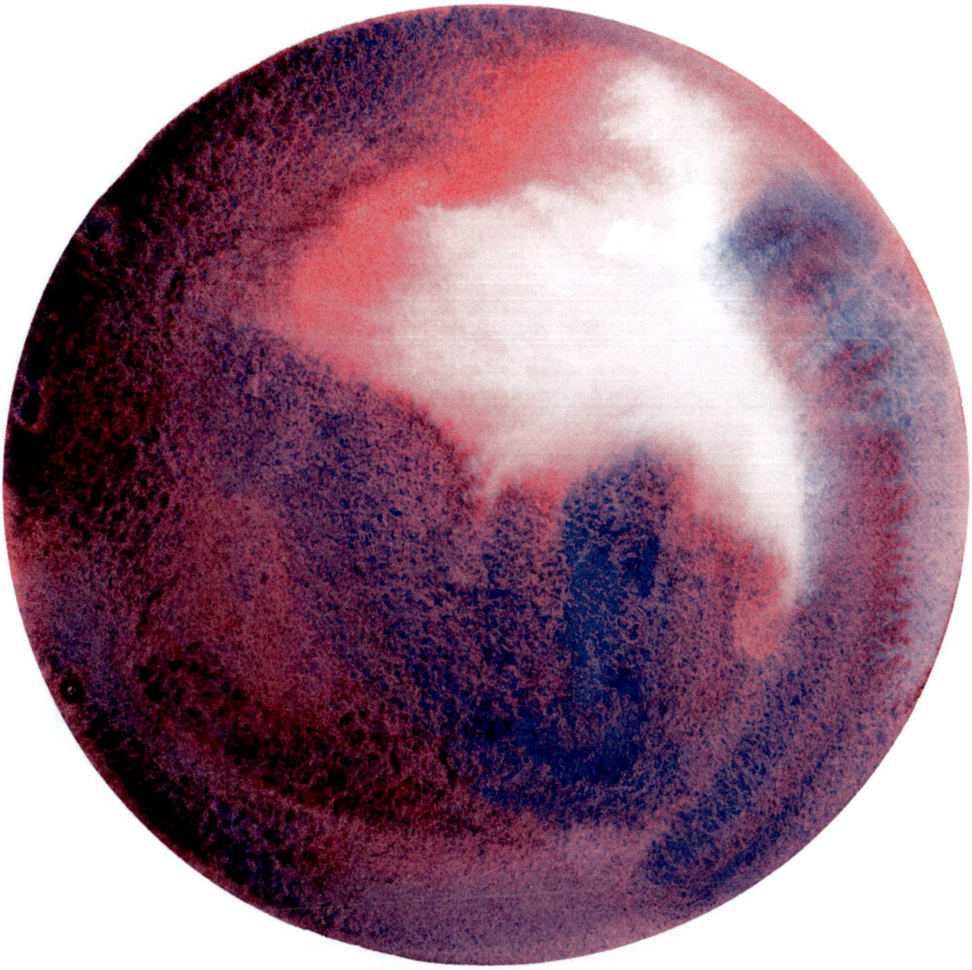

Fr

87 is francium

Half life of 22 minutes

Super radioactive

No weighable sample exists

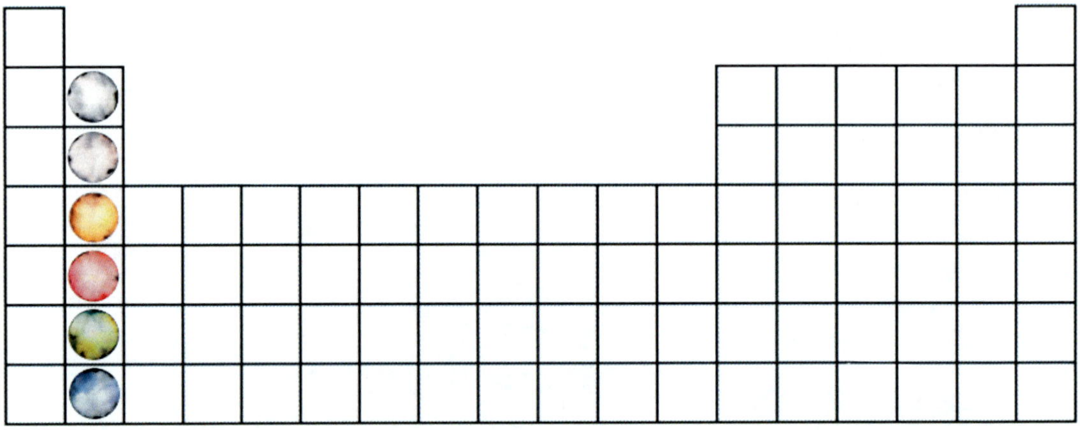

Group Two

Alkaline Earth Metals react with water

They always form plus two ions

Easy to spot (they're silver-white metals)

You can name three if you Be CalM

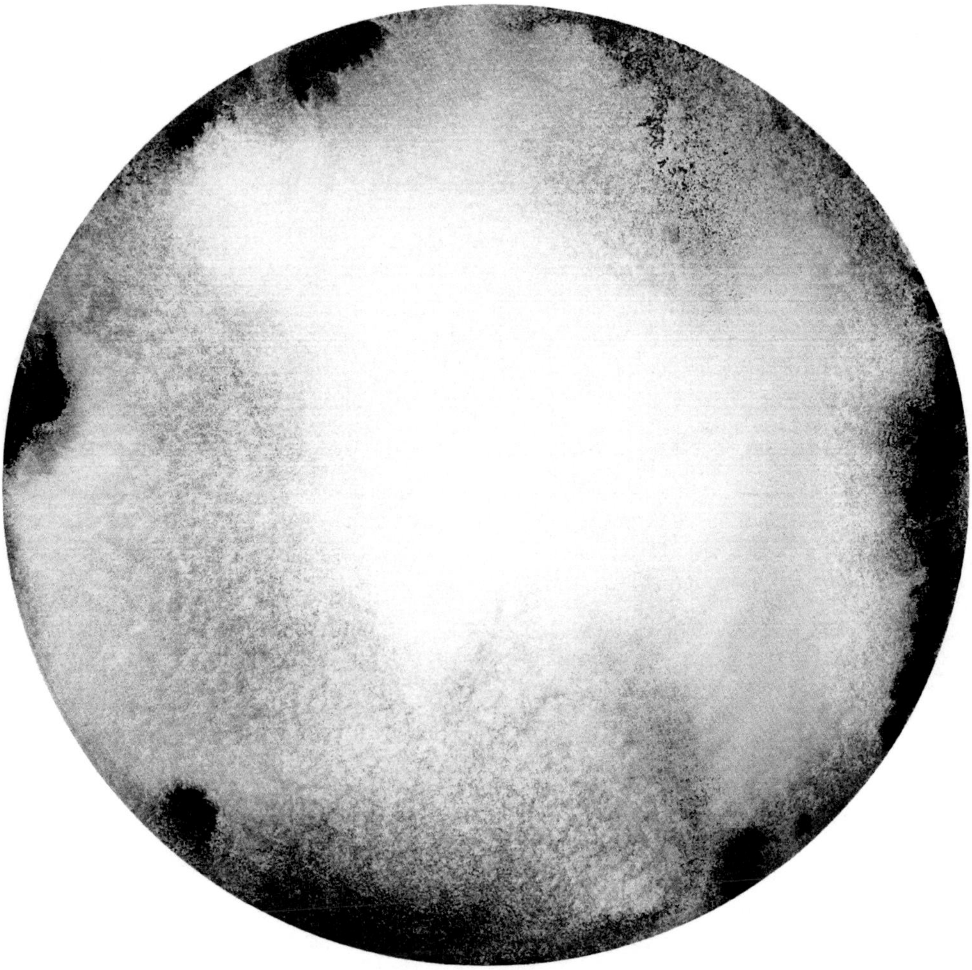

Be

Beryllium 4 is strong and light
Expensive and hard to corrode
In golf clubs and aerodynamics
It’s poison and in precious stones

Group 2

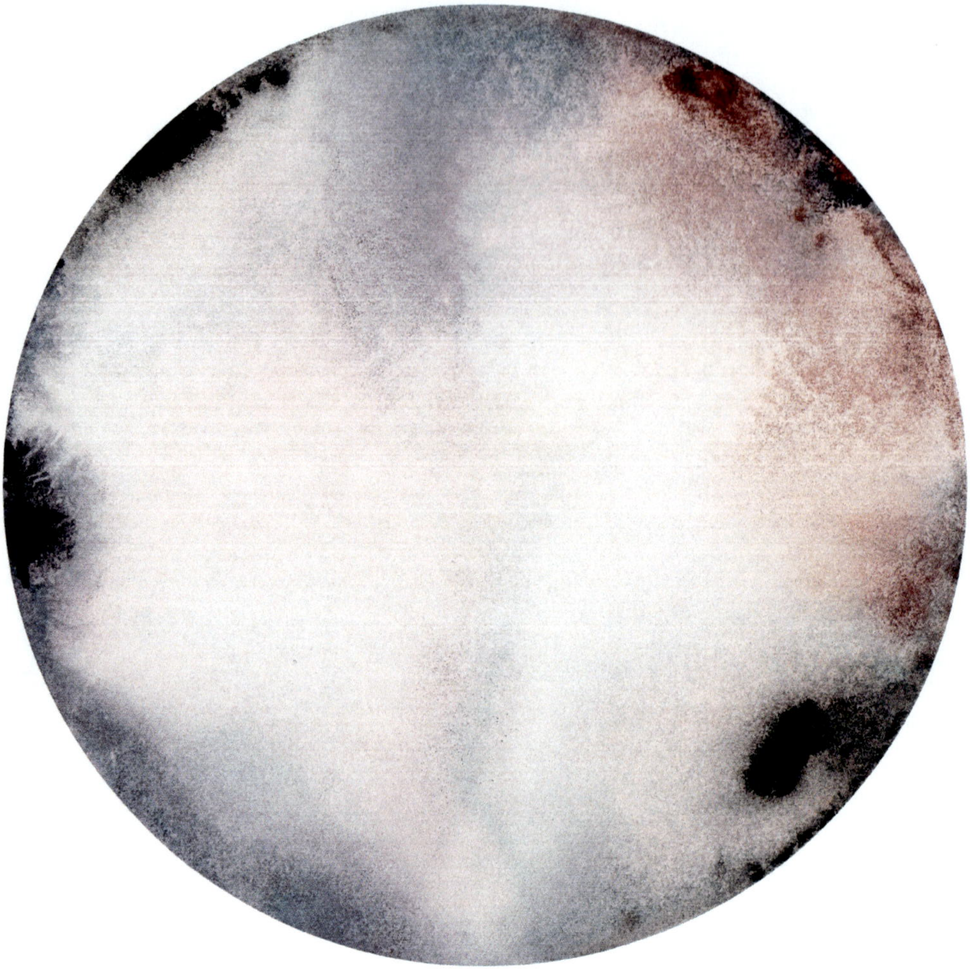

Mg

Number 12 is magnesium
Sourced from seas, it helps build your bones
In bombs, and flashbulbs, and alloys
Drink it when you just can't go

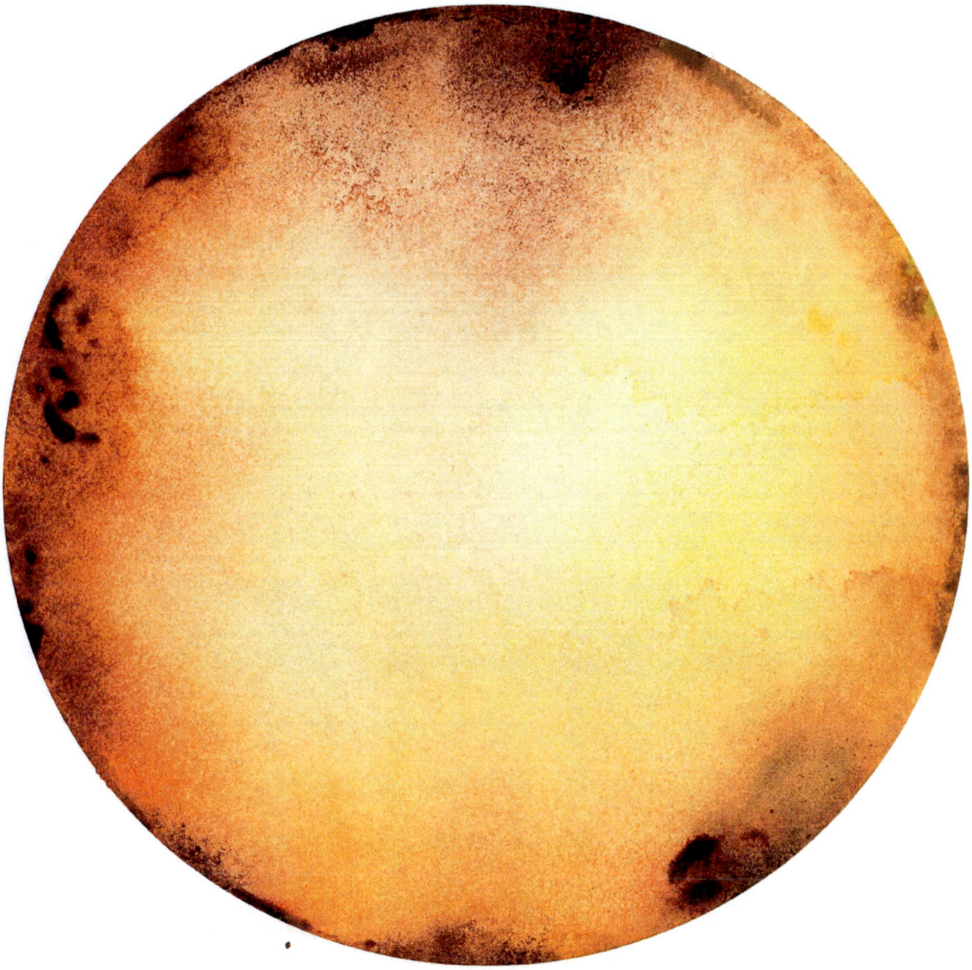

Ca

One kilogram of you is calcium
It's abundant (and number 20)
It loves to bond with nonmetals
It's Tums for your upset tummy

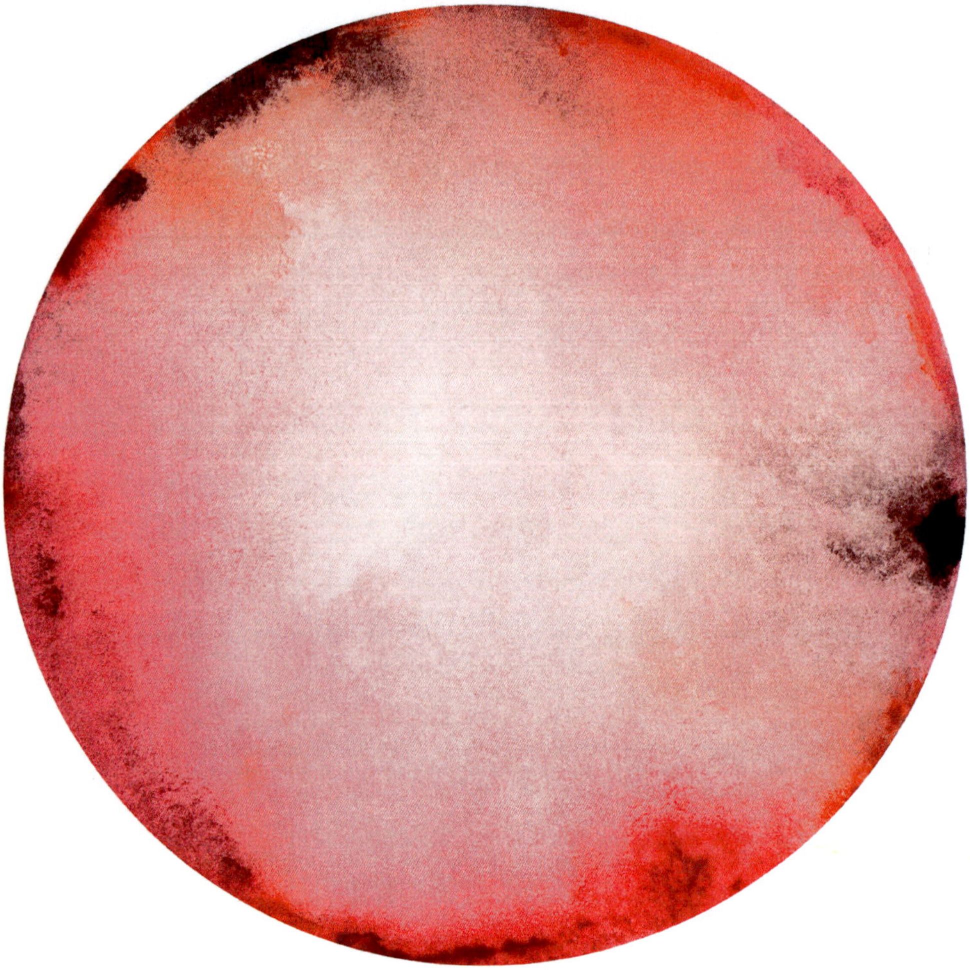

Sr

Strontium-90's from nuclear fallout
But regular strontium's fine
38 makes bright red fireworks
Active ingredient in Sensodyne

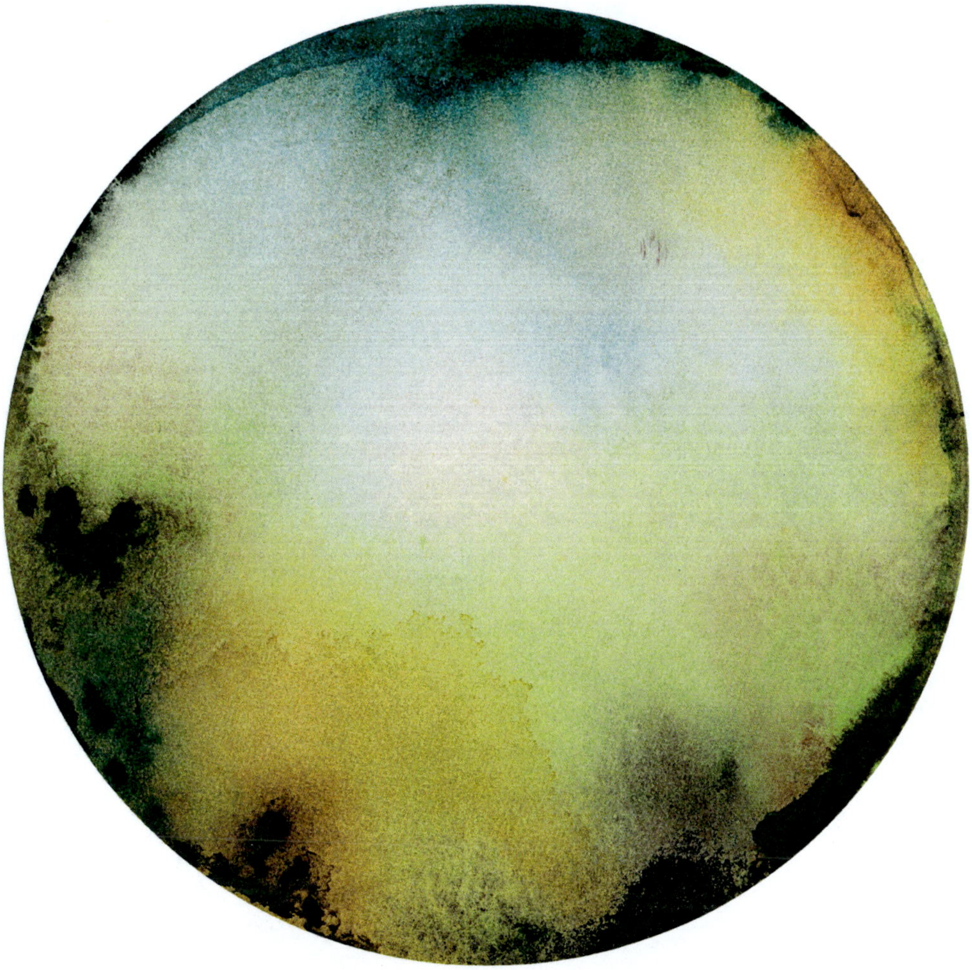

Ba

Drill for oil with 56
It's Greek for "heavy" (it's not)
"Getters" consume the oxygen
Barium enemas make things pop

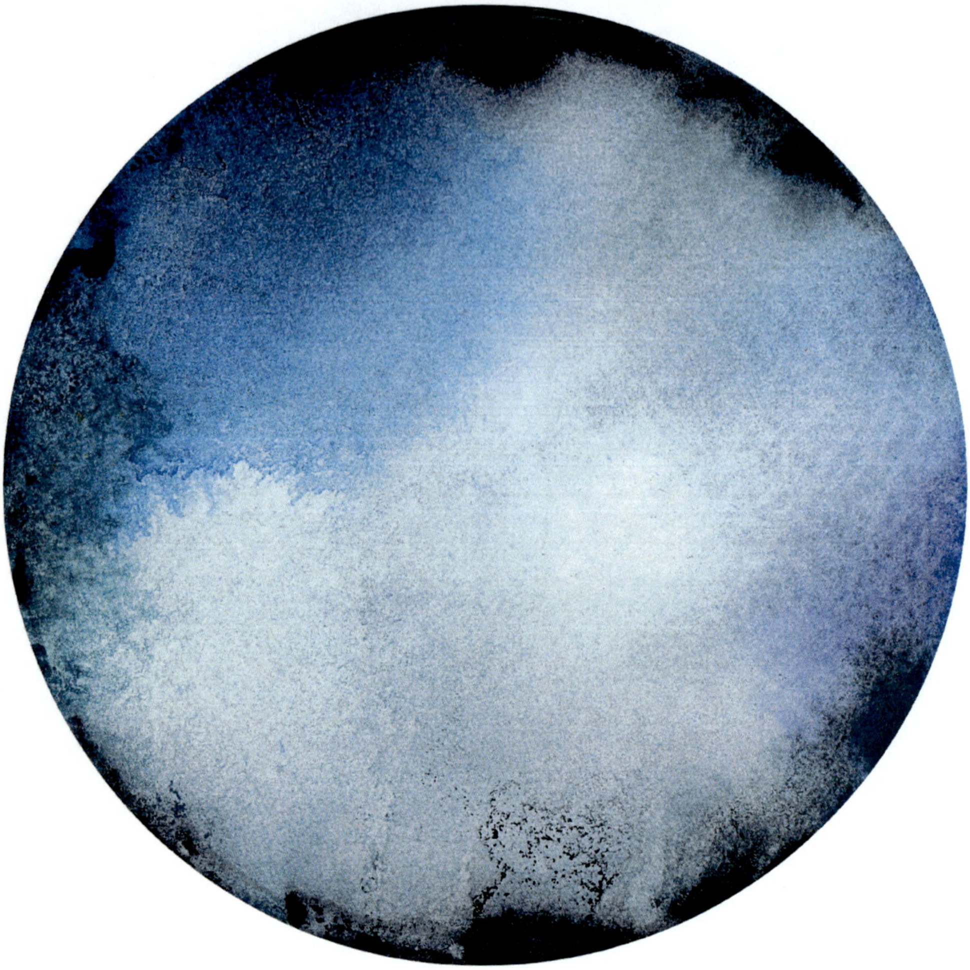

Ra

Radium 88

Creates a blue-ish glow

Once thought to be healthy

The Radium Girls say no

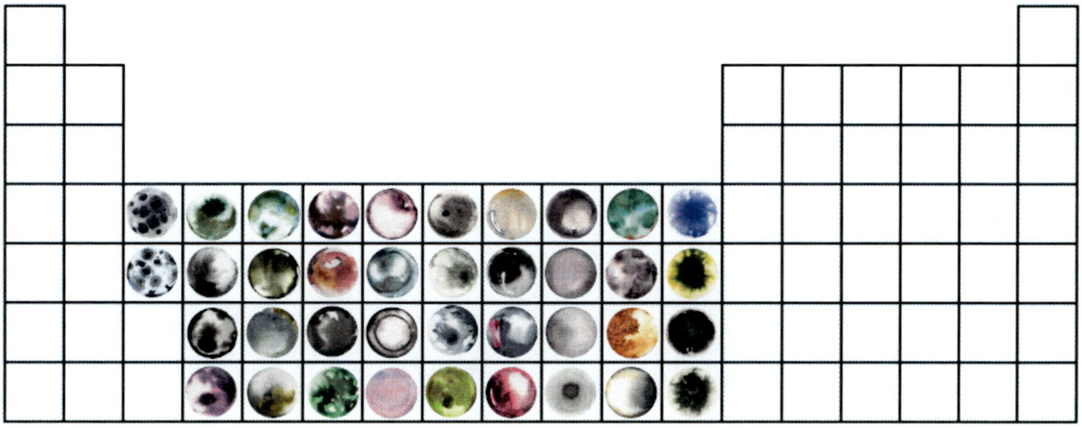

Transition Metals

D-Block Elements are partially filled

High boiling points and good conductors

With multiple oxidation states

They form bonds and mix well with others

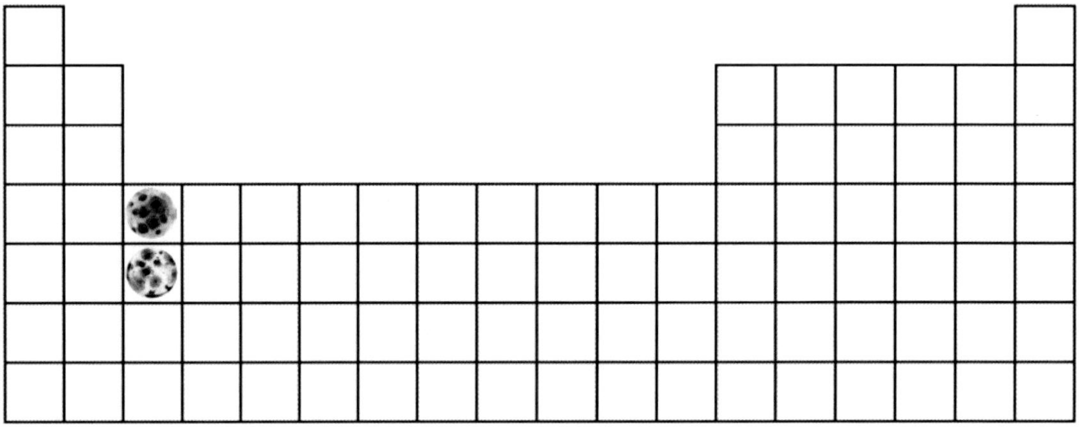

Group Three

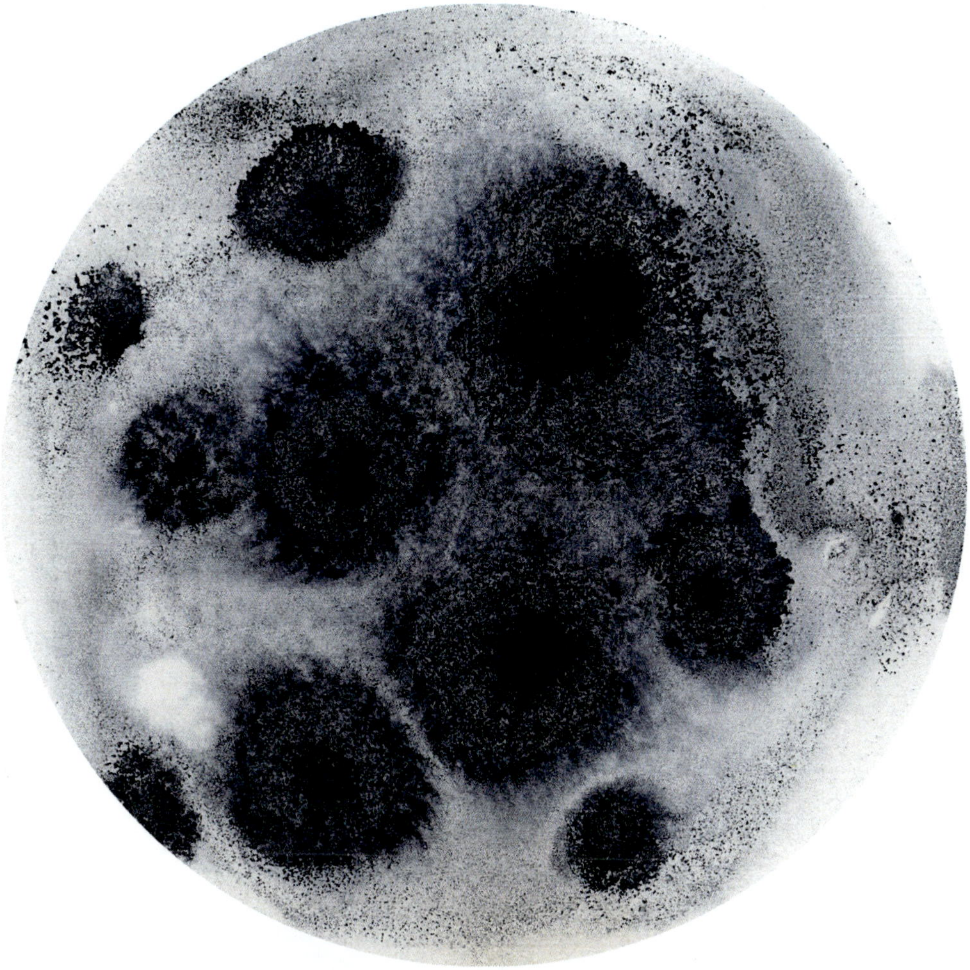

Sc

Scandium 21 costs a lot to source
You can find it in fancy bike frames
It wants to star in renewable energy
In movies and at baseball games

Group 3

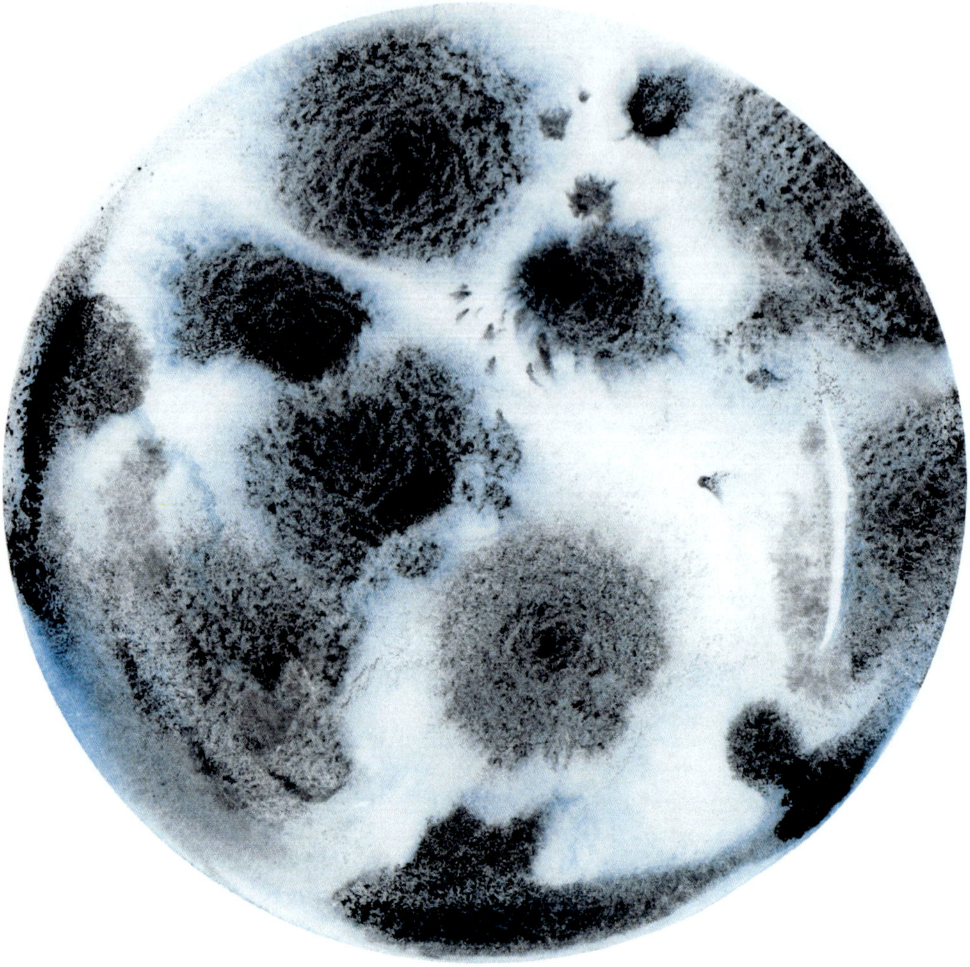

Y

Your "diamond's" sparkle is yttrium

One stable isotope in nature

Causes lung disease, fights cancer

39 makes a pulsed laser

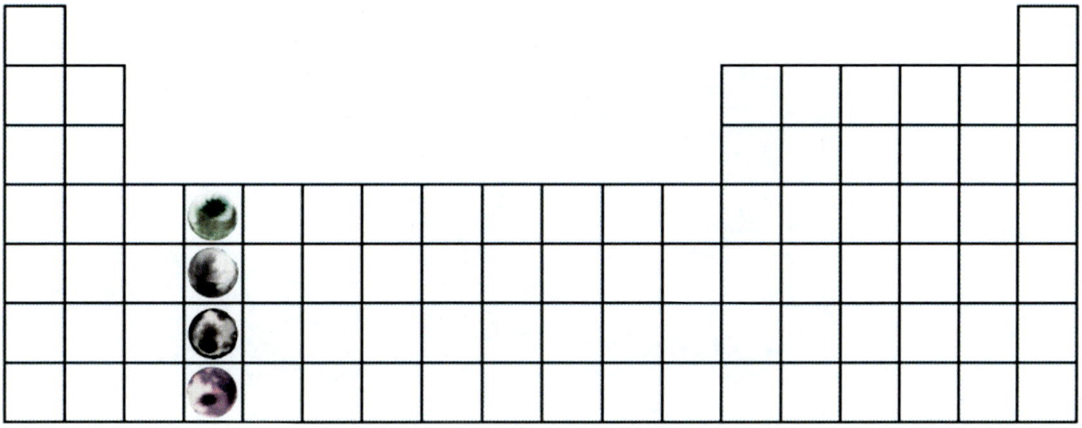

Group Four

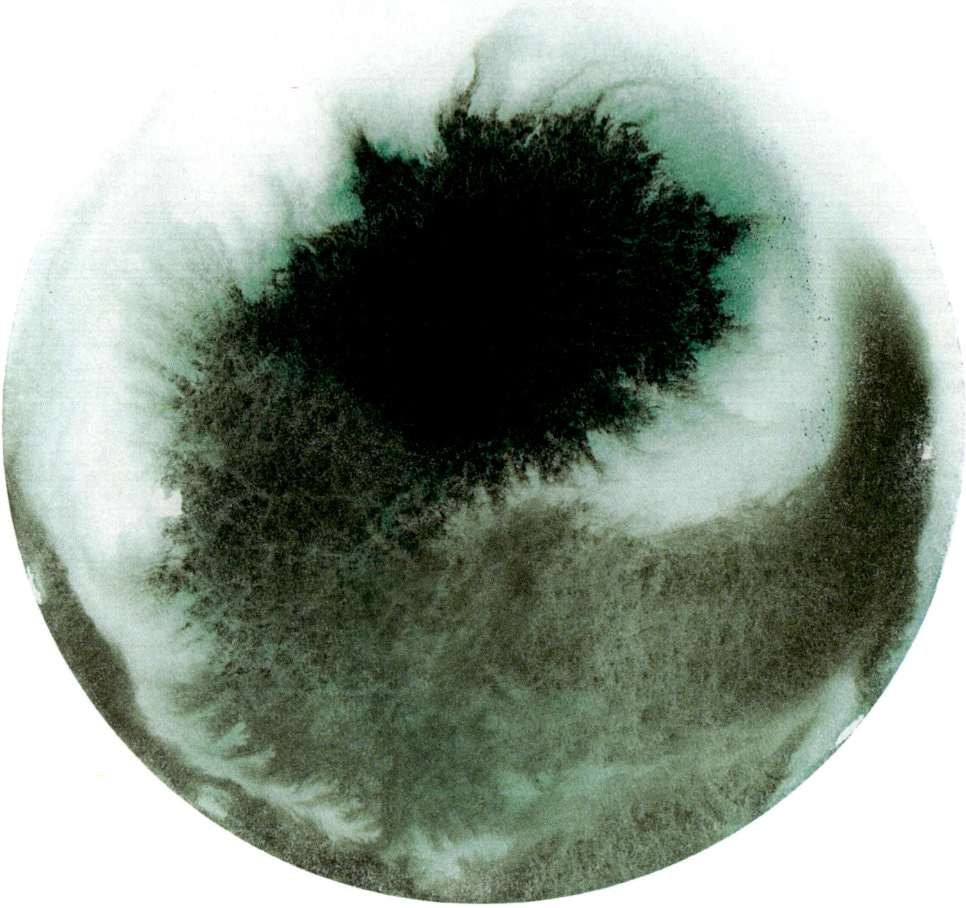

Ti

Titanium is element 22

Body implants and makes things opaque

Strong and 45% lighter than steel

So popular some people fake it

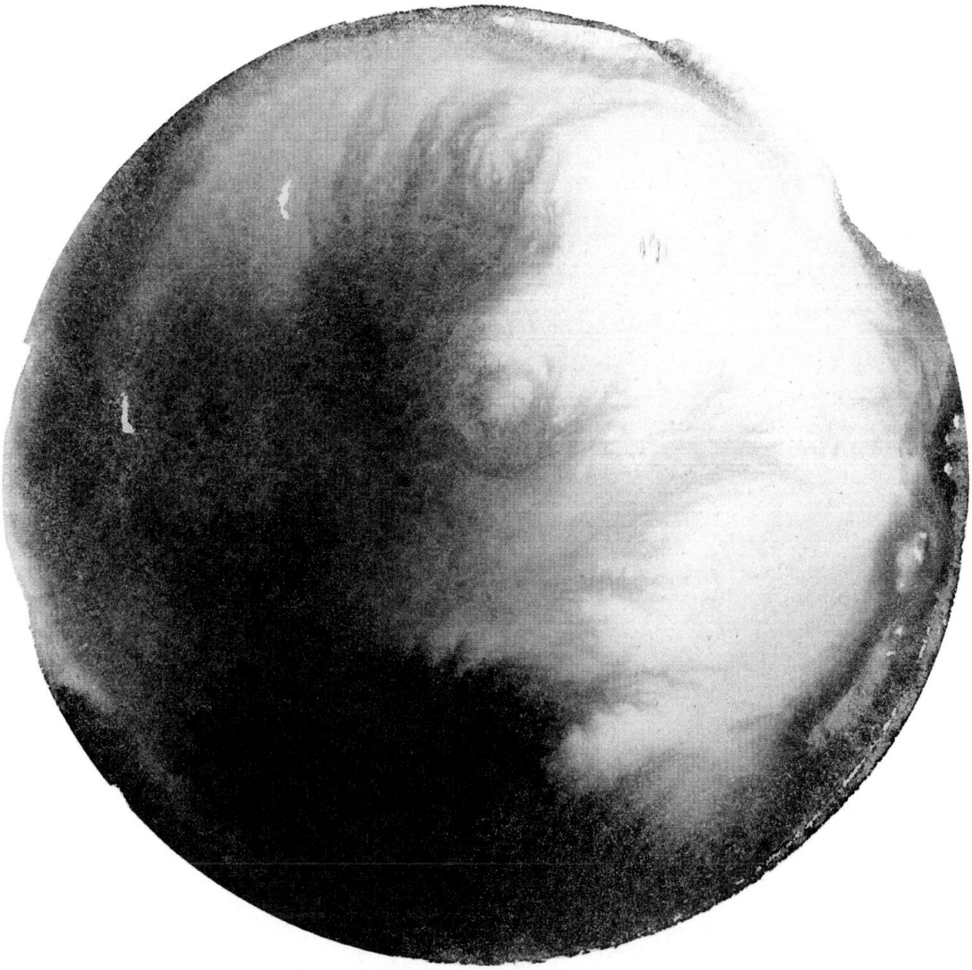

Zr

Zirconium is molten metal's ladle
Few metals can shield heat and bend
Neutron transparency and grinders
Number 40 is a girl's best friend

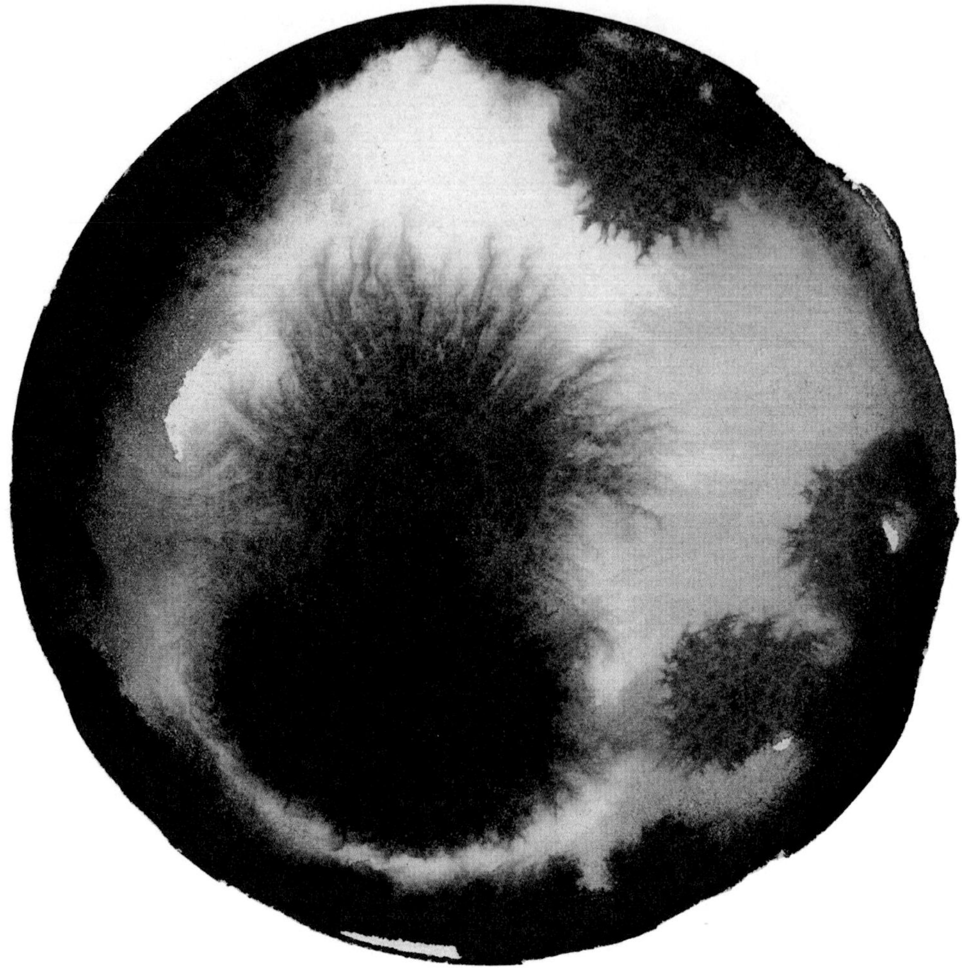

Hf

Hafnium halts a nuclear reaction
And makes perfect plasma torch tips
72 has an elemental twin
High-k in your phone's microchip

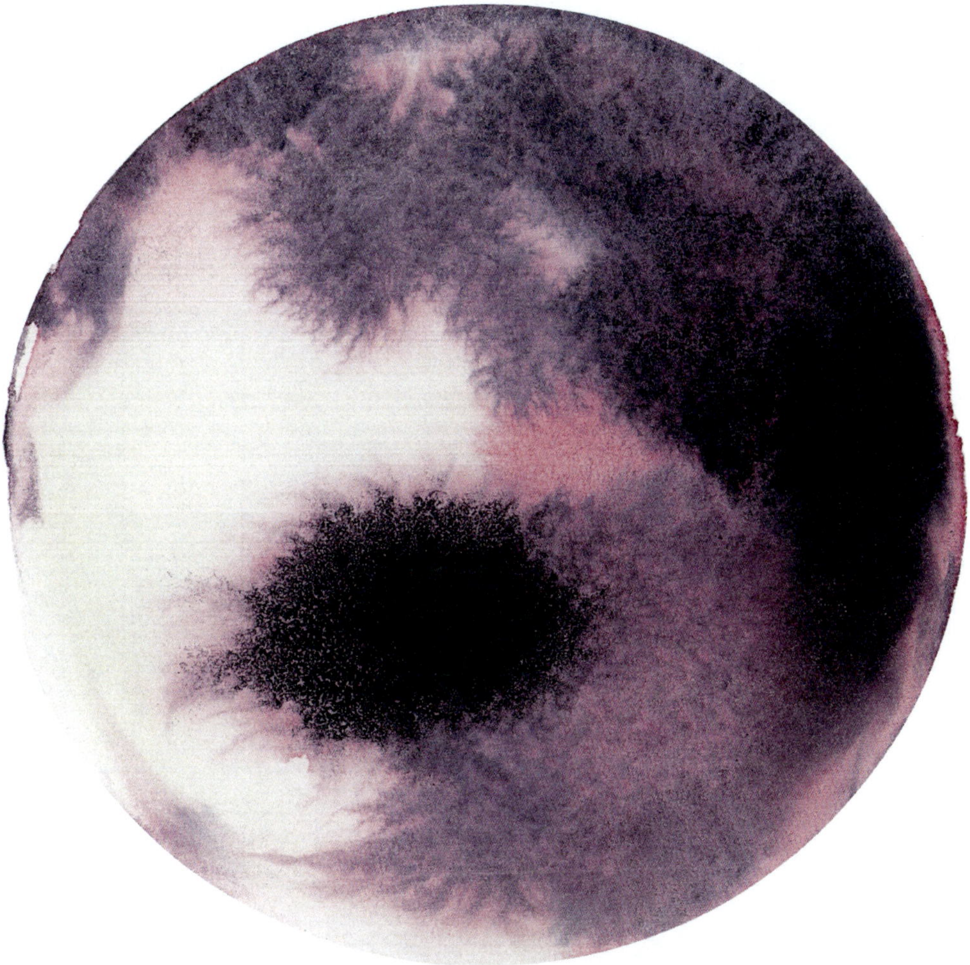

Rf

Synthetic rutherfordium 104

If you can see it, there's enough to kill you

A decades-long war over picking its name

Not much known, so research continues

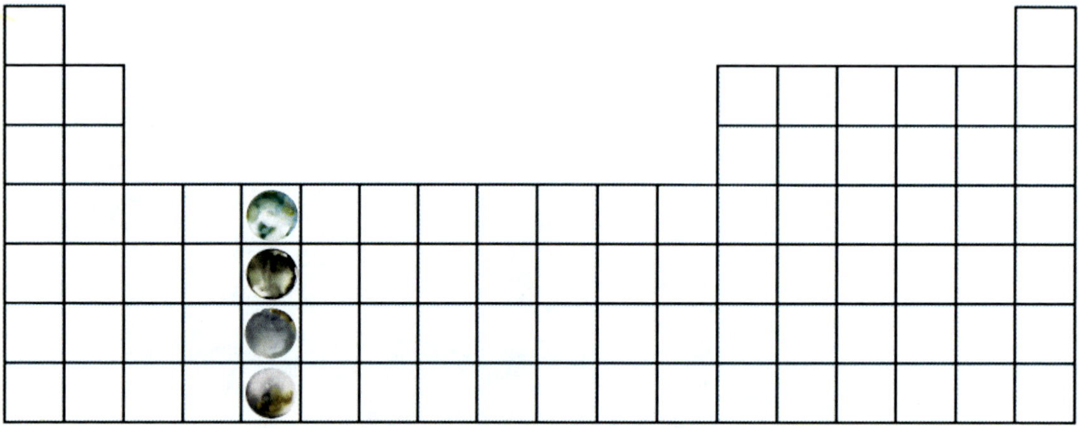

Group Five

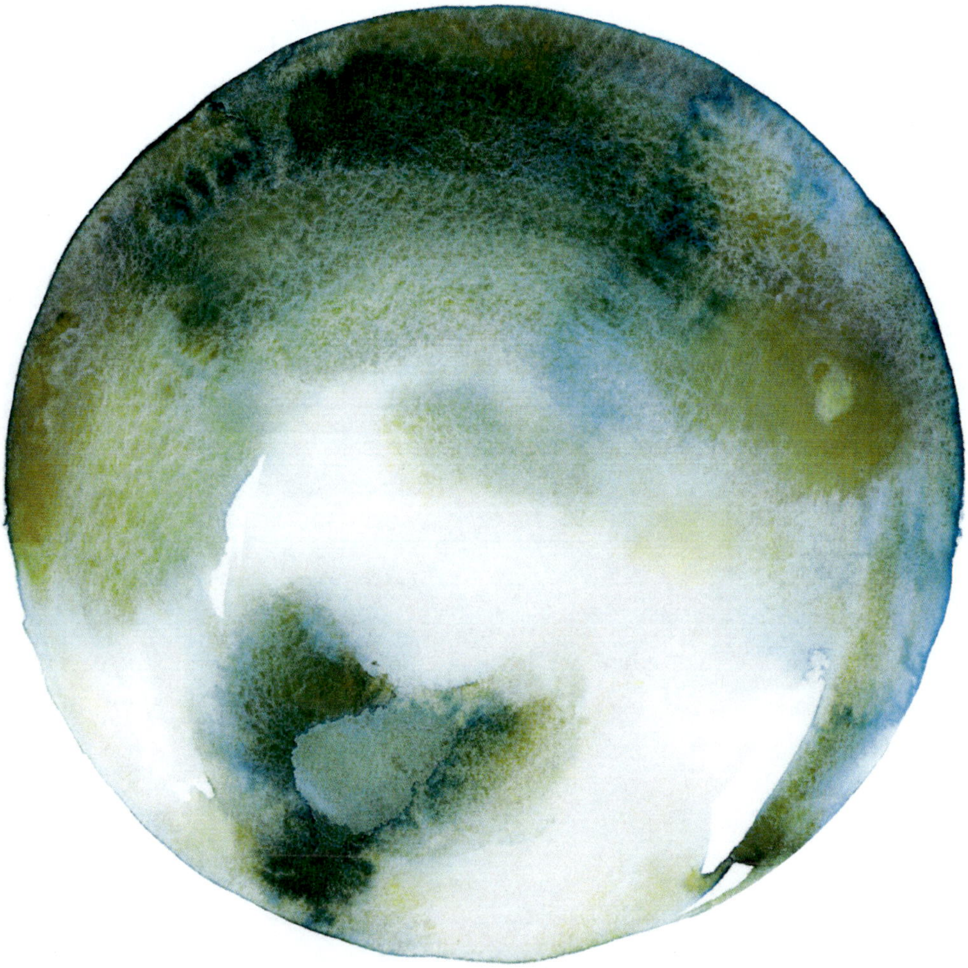

V

23 is much harder than steel

80% goes into High Speed

In batteries and reactors

Vanadium makes some emeralds green

Group 5

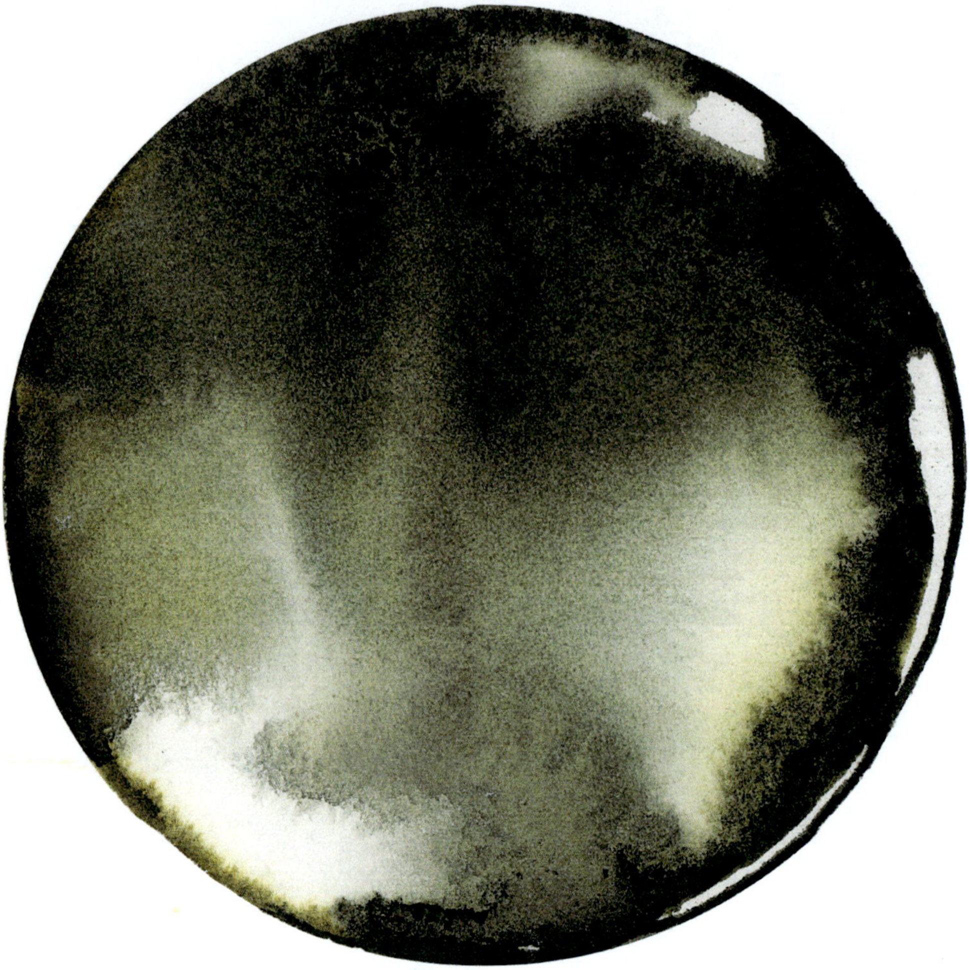

Nb

Niobium number 41

Magnets in MRI machines

Meteorites and beautiful oxides

Rocket nozzles and belly rings

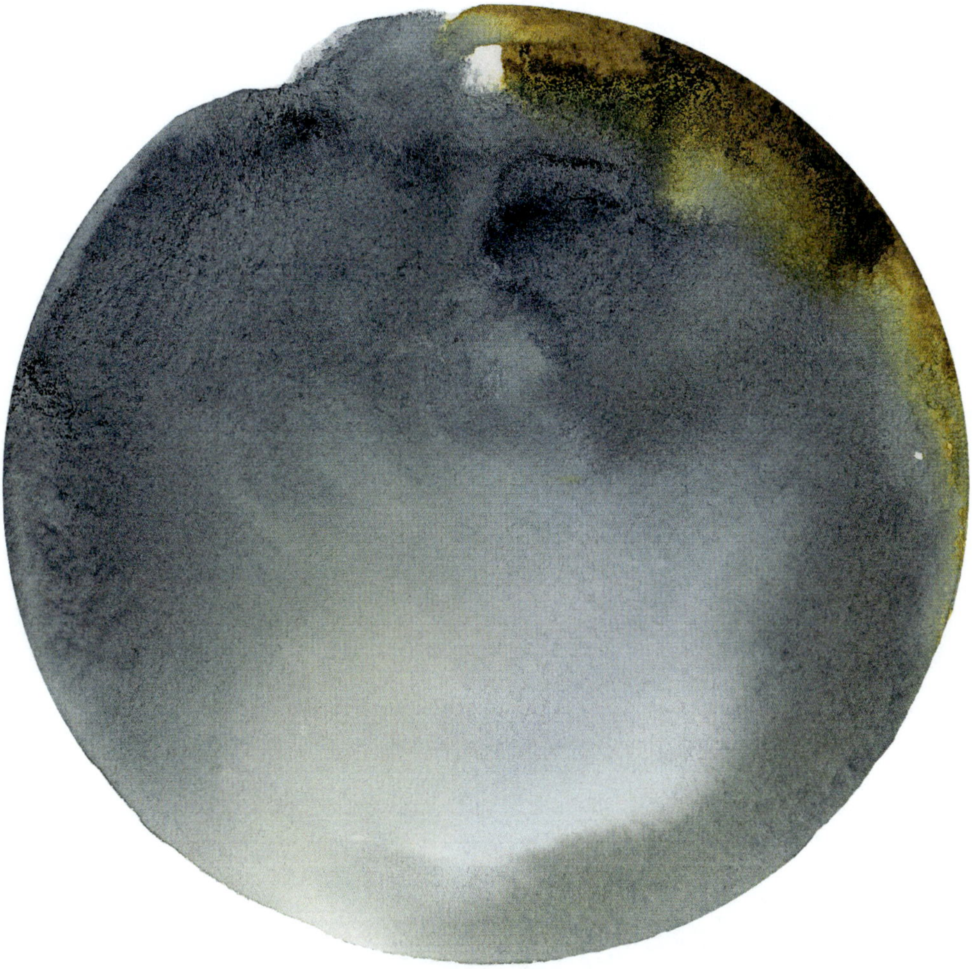

Ta

73: Tantalum prosthetics
Body armor contains its ceramics
Conflict mineral inside your Xbox
Was the light aboard the Titanic

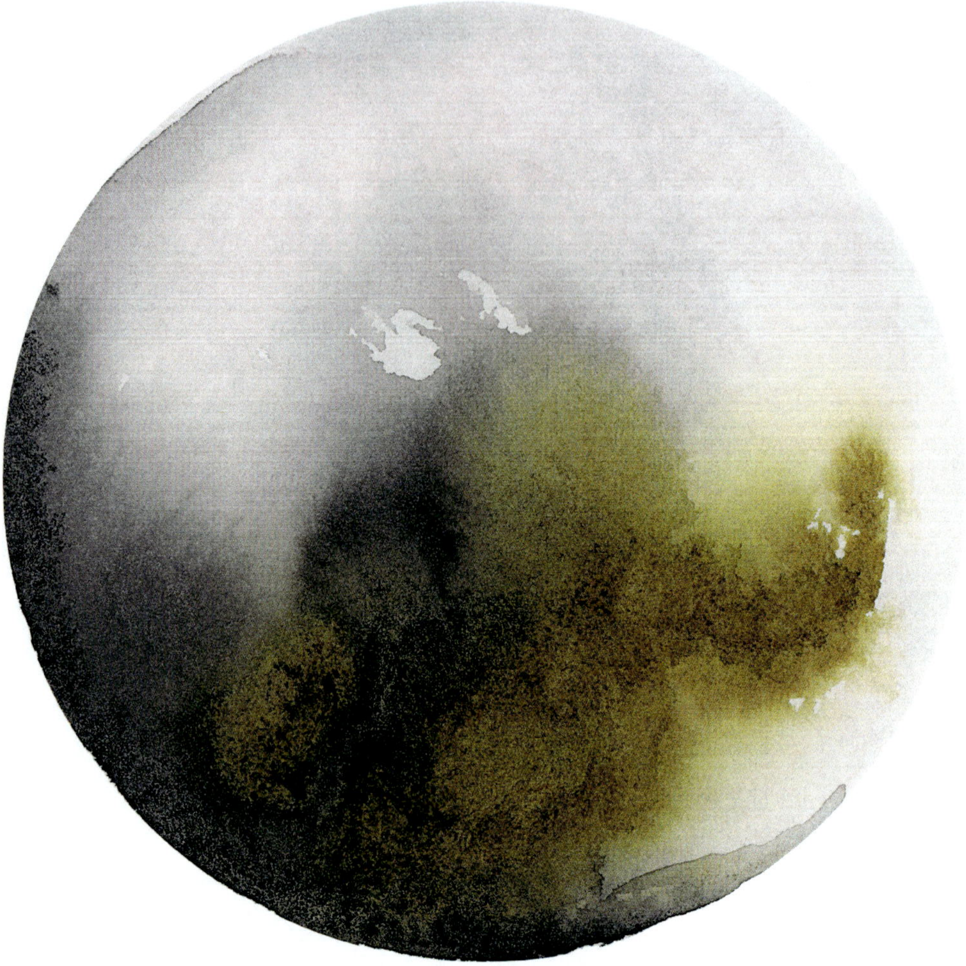

Db

Lab-made dubnium, number 105
An old timer - lasts more than a day!
Named after the lab that found it
Longest life of "super heavyweights"

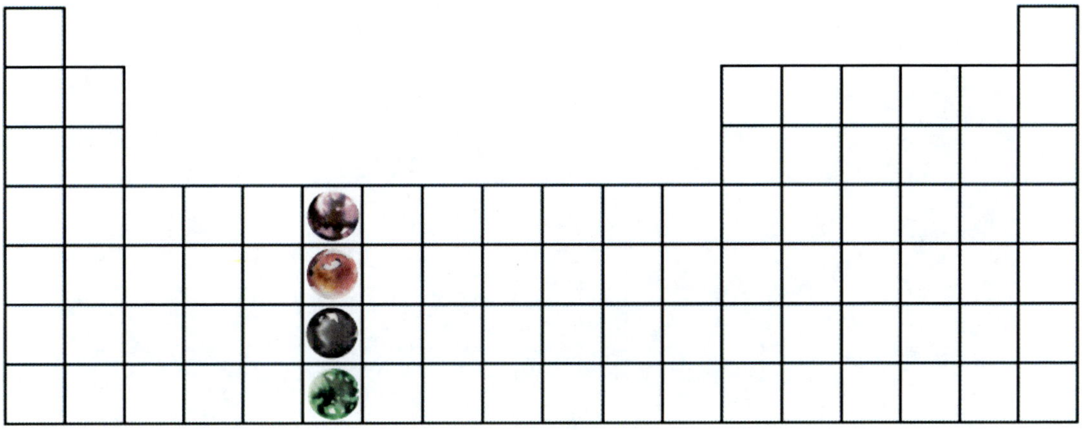

Group Six

Group 6

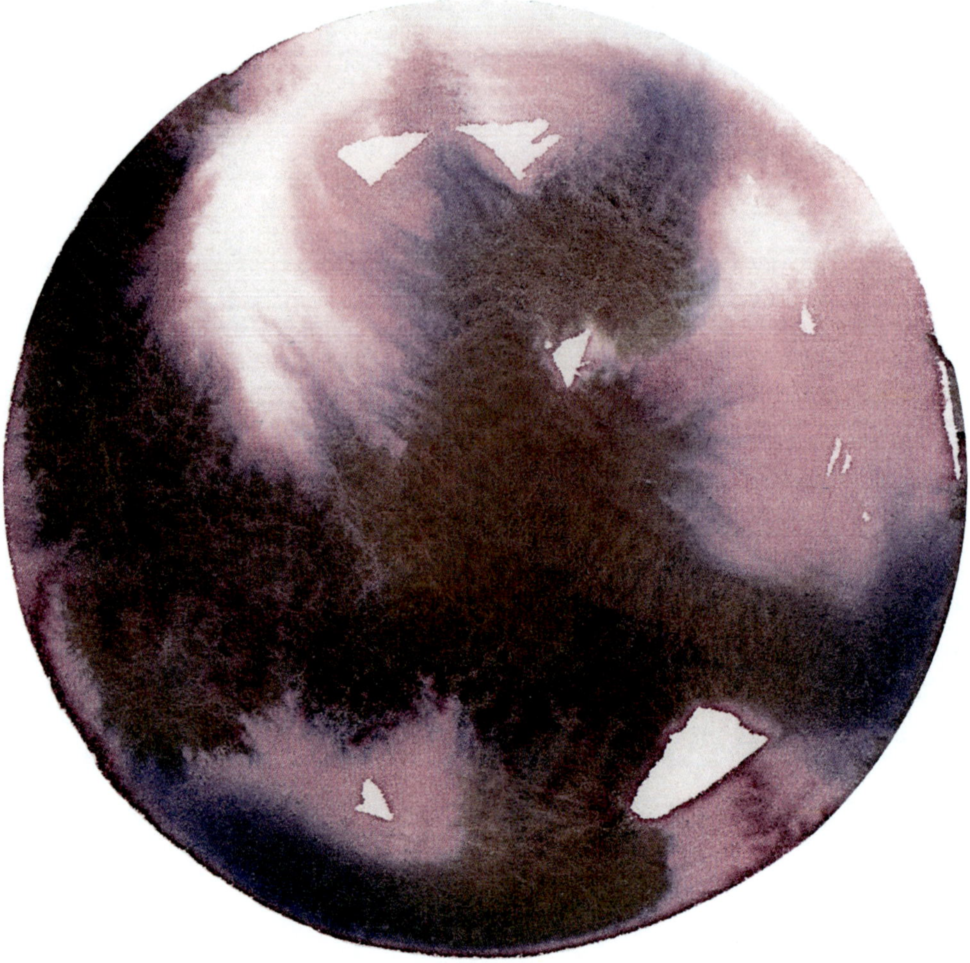

Cr

24 chromium plates your forks
Its colors are lustrous and rich
One quarter of stainless steel alloys
Can be toxic (so says Brockovich)

Group 6

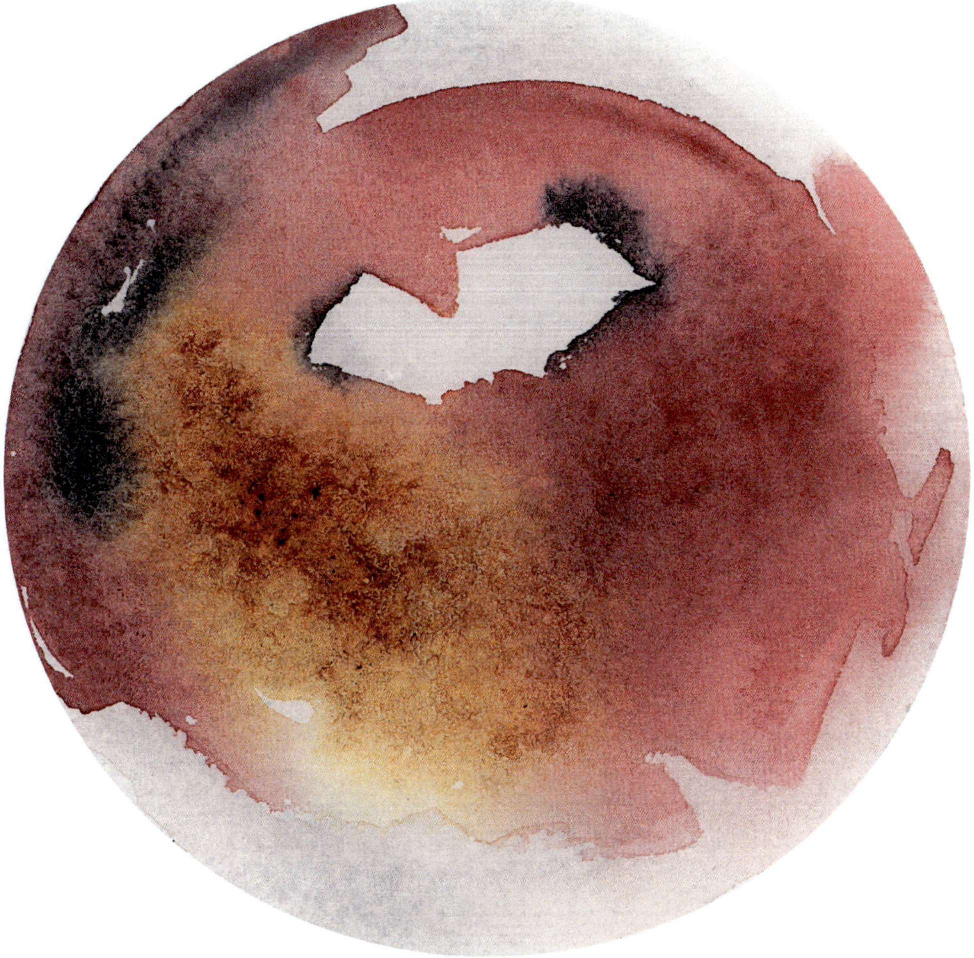

Mo

Molybdenum 42
Withstands a thermal blast
It fills the moly cow
And reveals the ocean's past

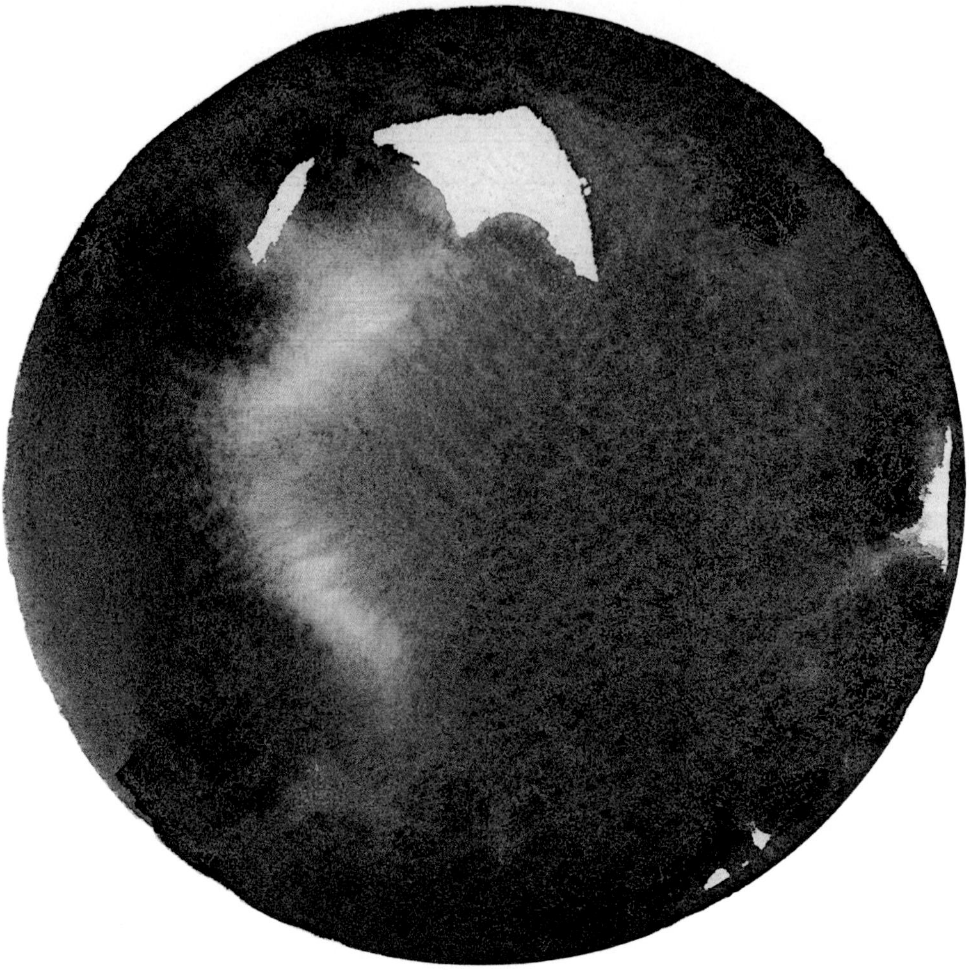

W

Cheap and tough tungsten, number 74

Cutting tools and bulletproof plating

Lightbulb filaments (90% wasted)

Smartphones and F1 counterweighting

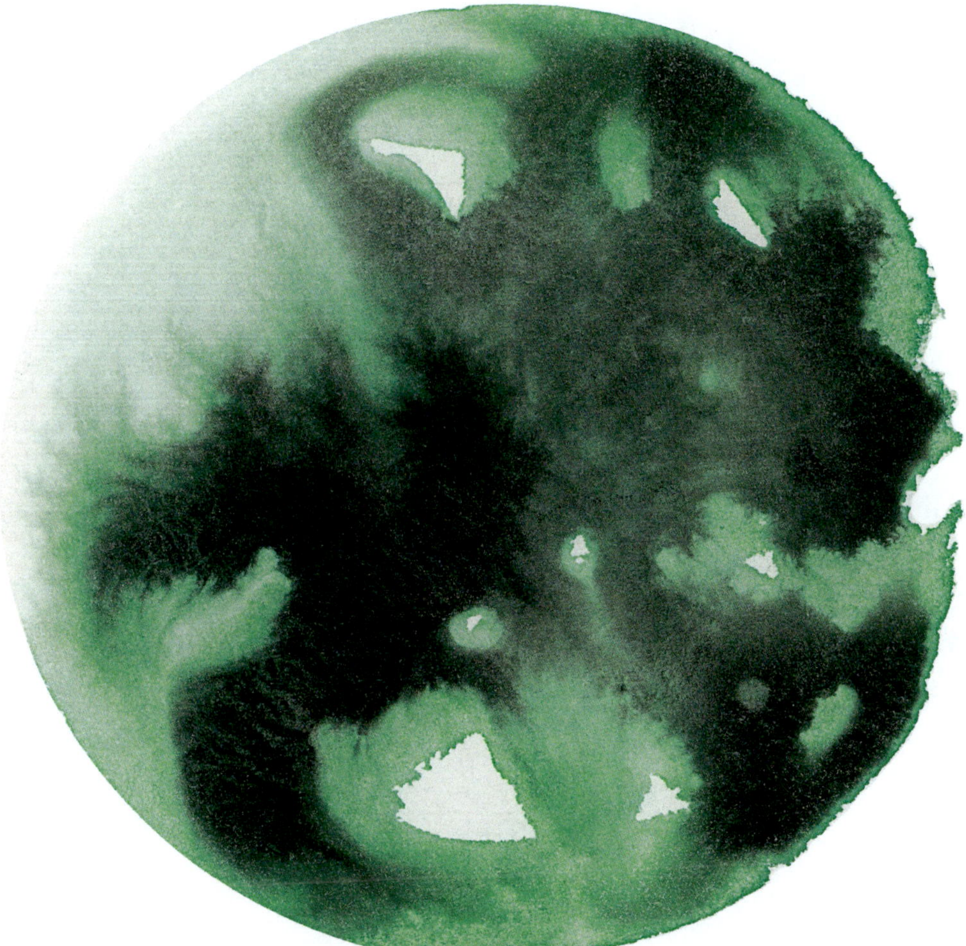

Sg

First element named for a living dude

Synthesized seaborgium 106

Radioactive and unstable

Tungsten's twin; homologous

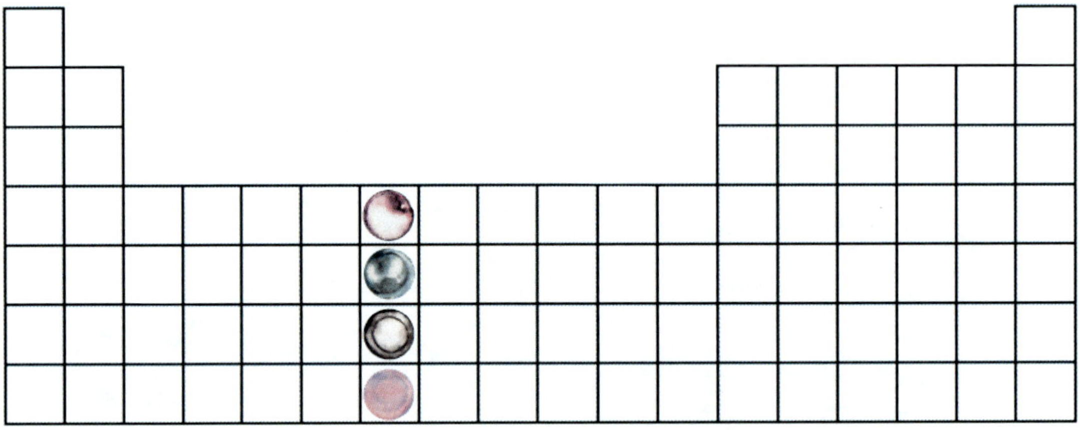

Group Seven

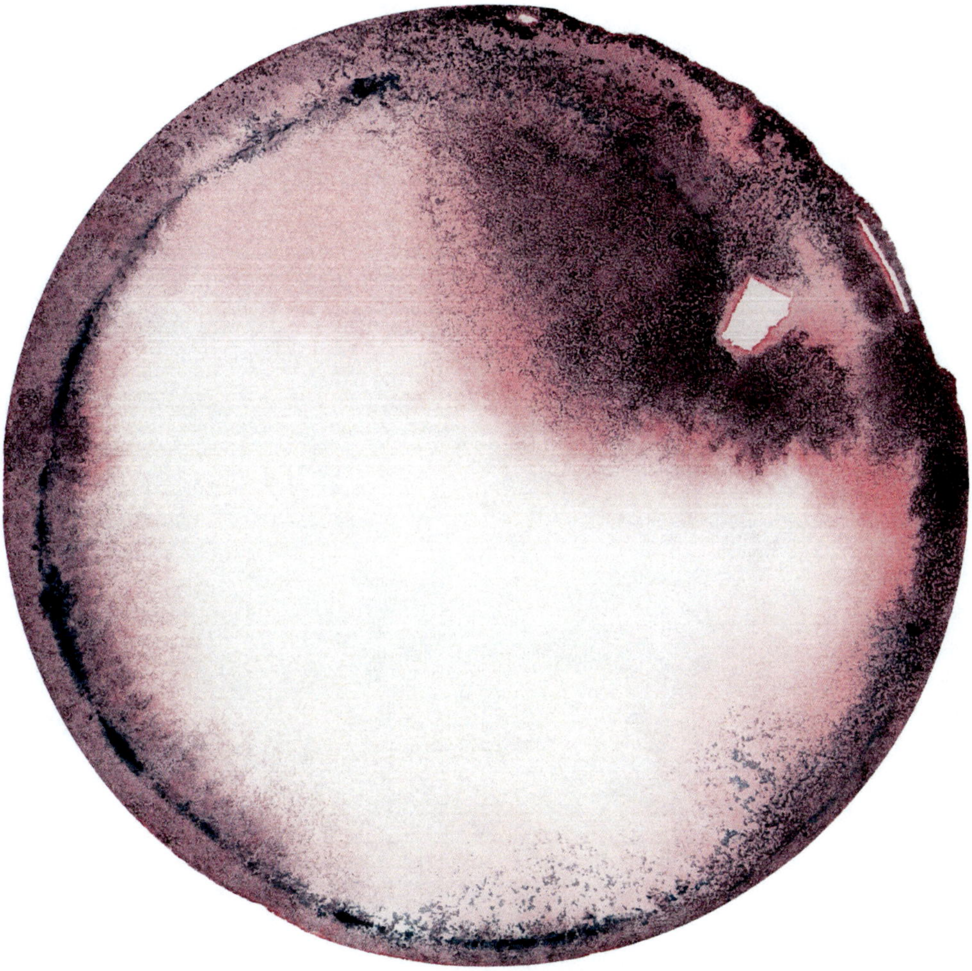

Mn

Number 25 found in cave paintings

It’s in steel, batteries, and cans

Exposure causes Manganese Madness

Project Azorian CIA scam

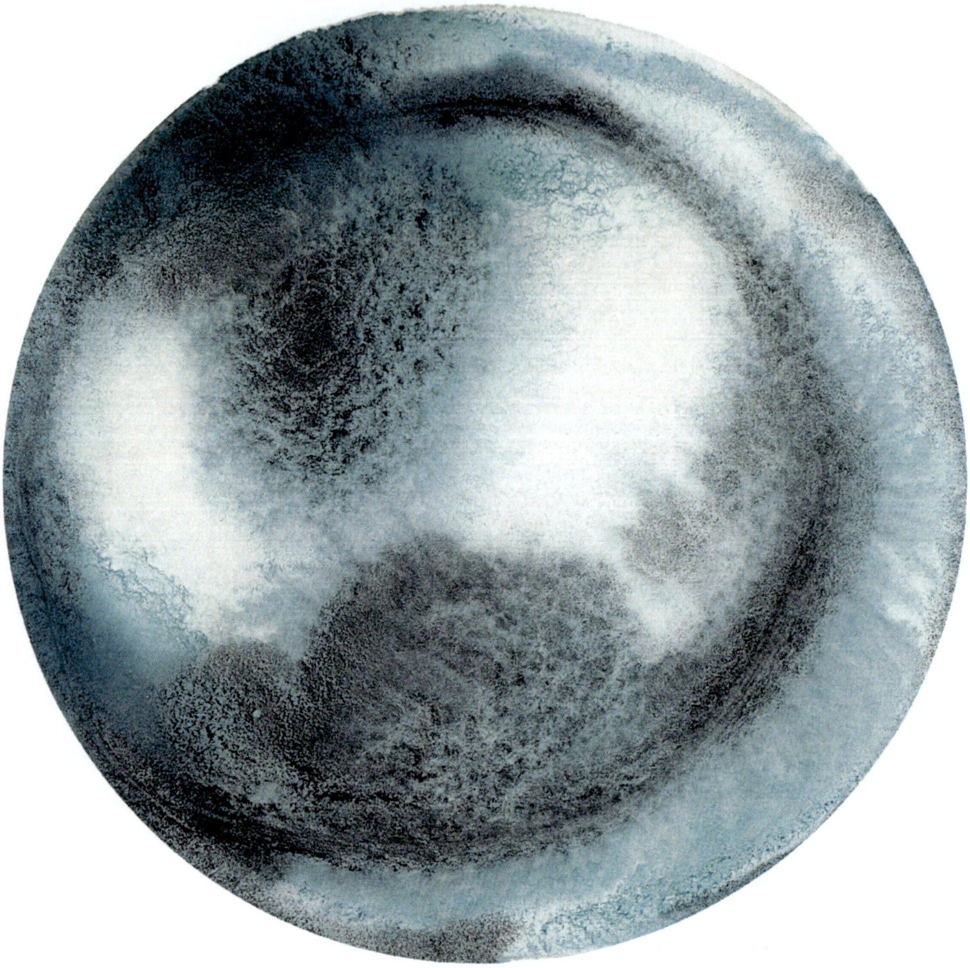

Tc

Technetium is number 43

First artificially made element

A radioactive bone seeker

Product of stellar development

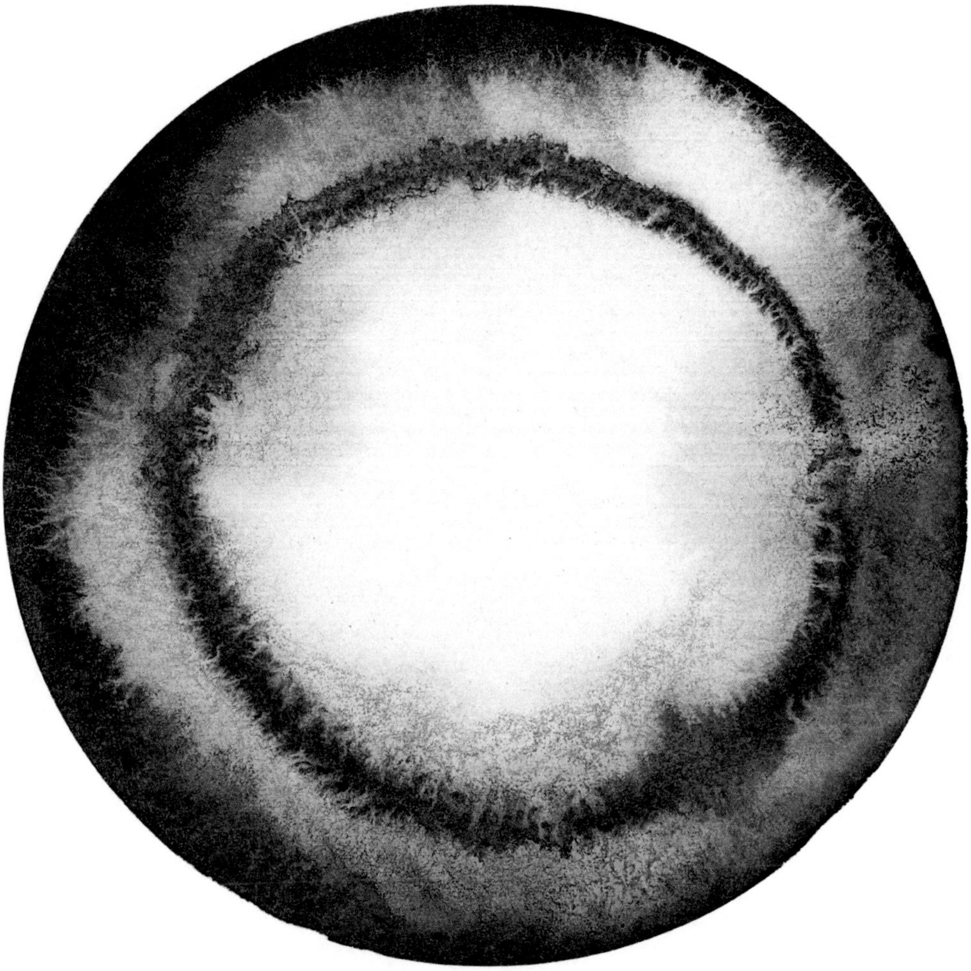

Re

Rhenium 75, forged in exploding stars
Strategic resource with heat resistance
Three quarters of use is fighter jet engines
Last stable element to reveal its existence

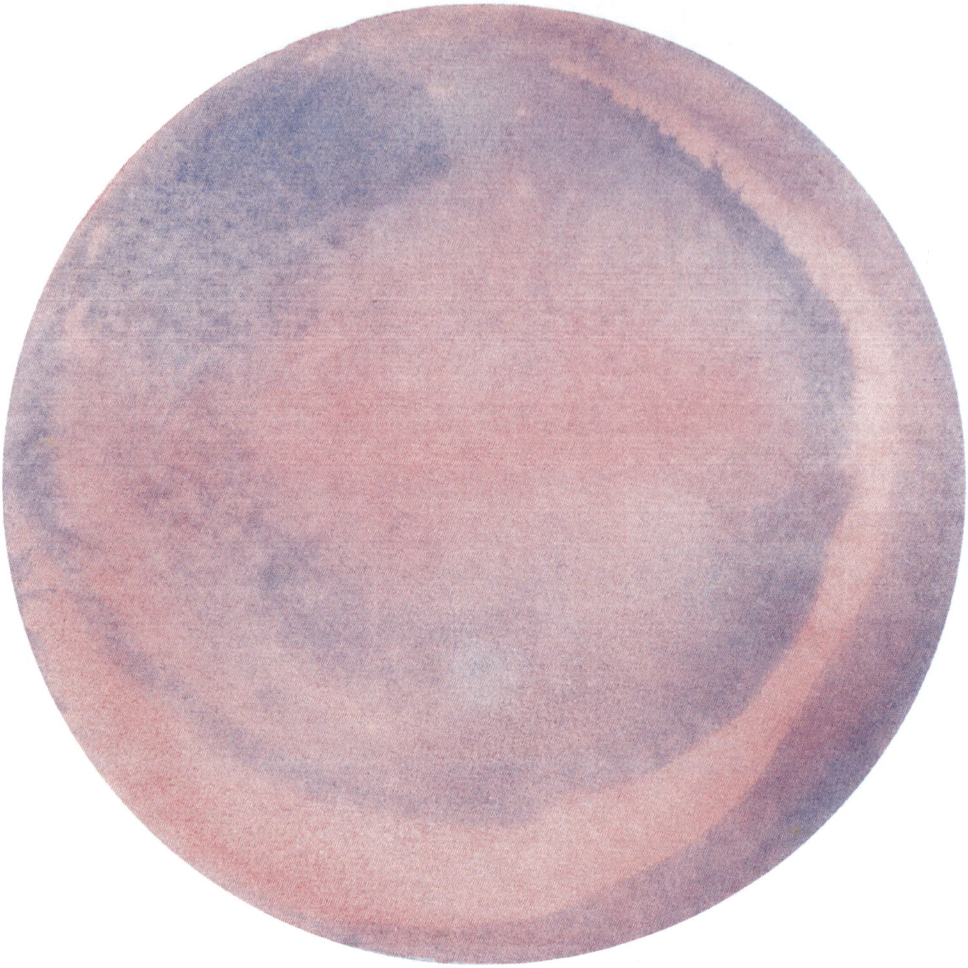

Bh

Lab-made bohrium number 107
Discovered at German institution
Expensive to make and decays real fast
It's produced through nuclear fusion

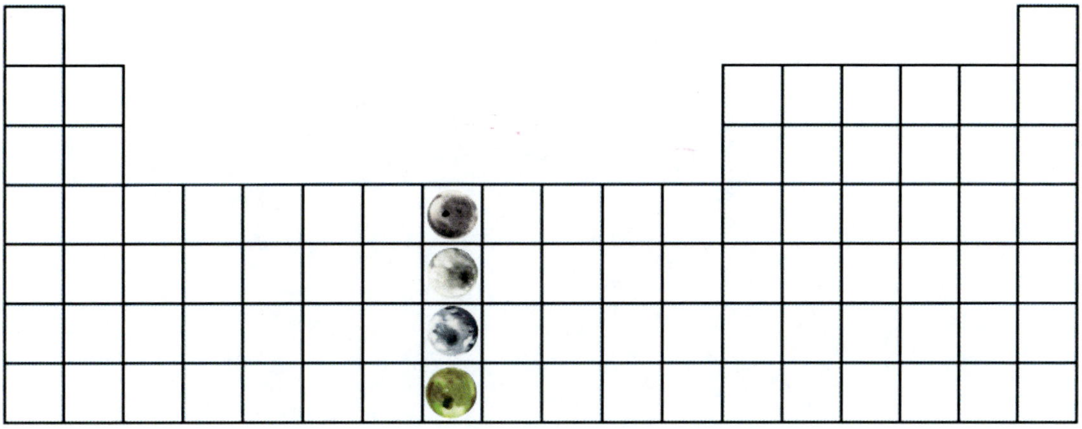

Group Eight

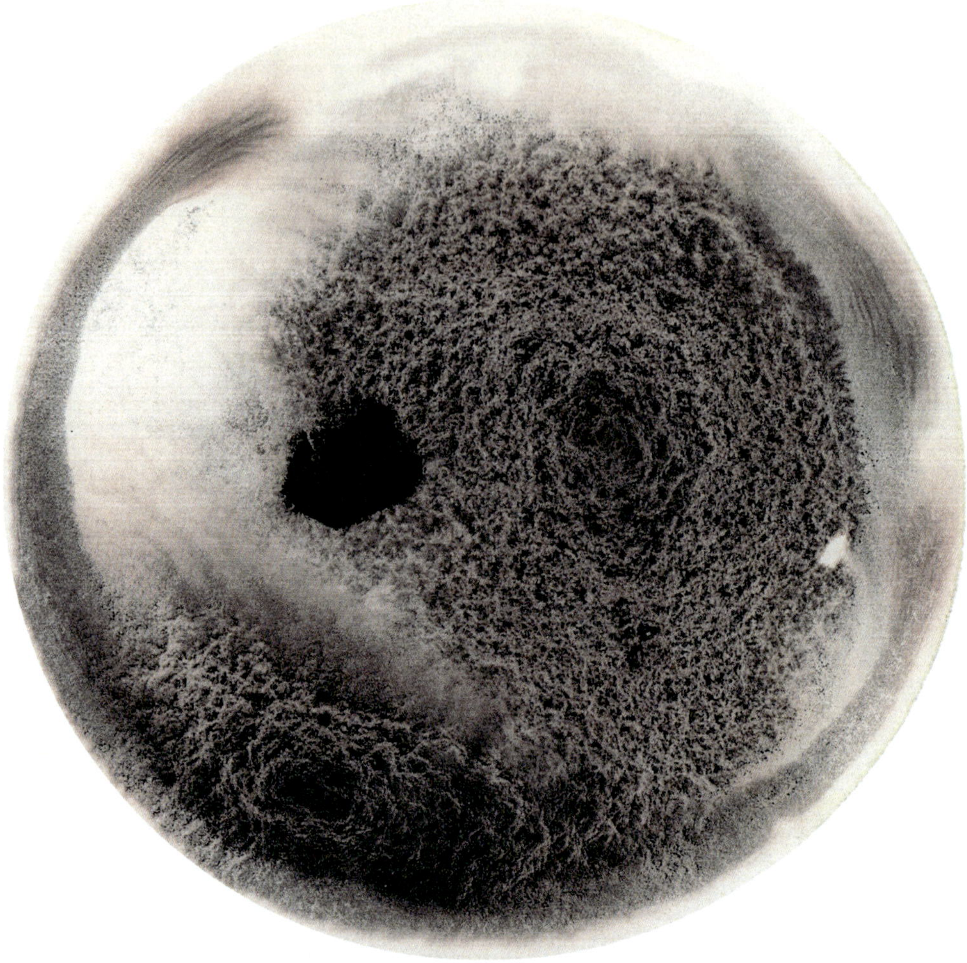

Fe

The Iron Age starts with element 26

The backbone of civilization

Found in cereal, blood, and meteorites

Scatters and absorbs radiation

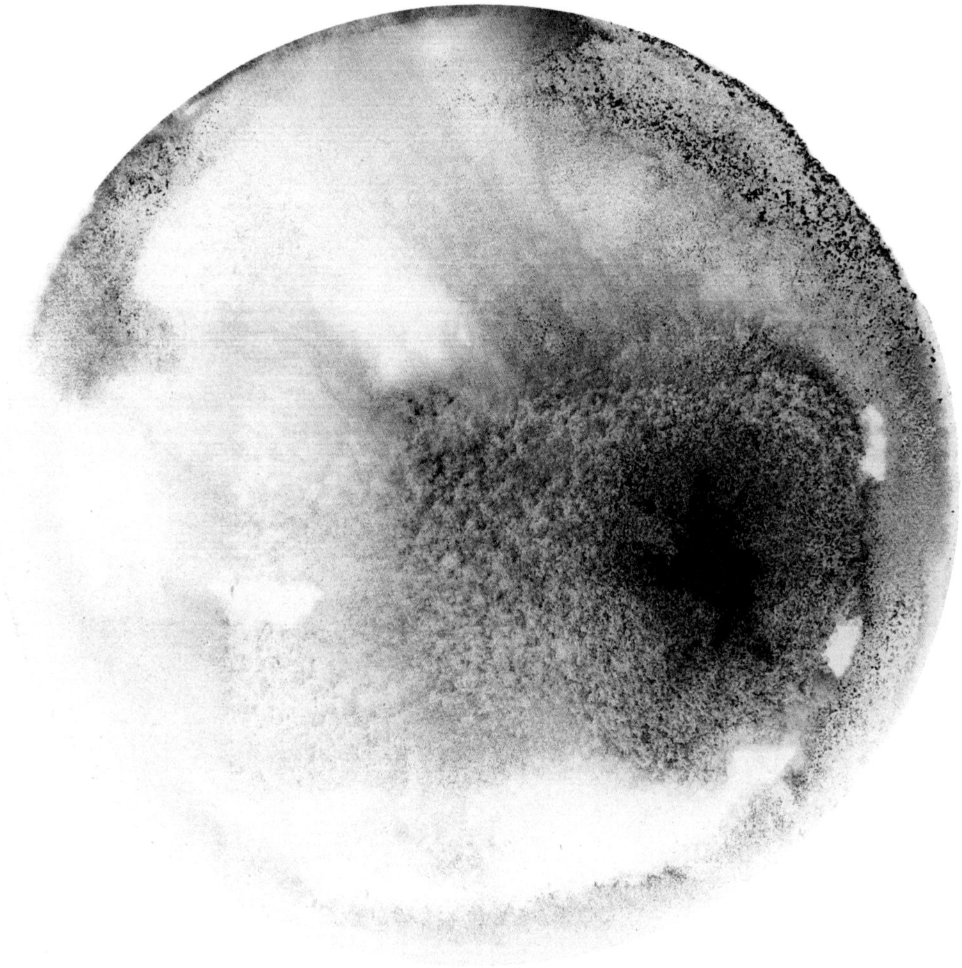

Ru

Ruthenium is 44
Makes high performance turbine blades
One hundred times more rare than gold
Present in outer space rays

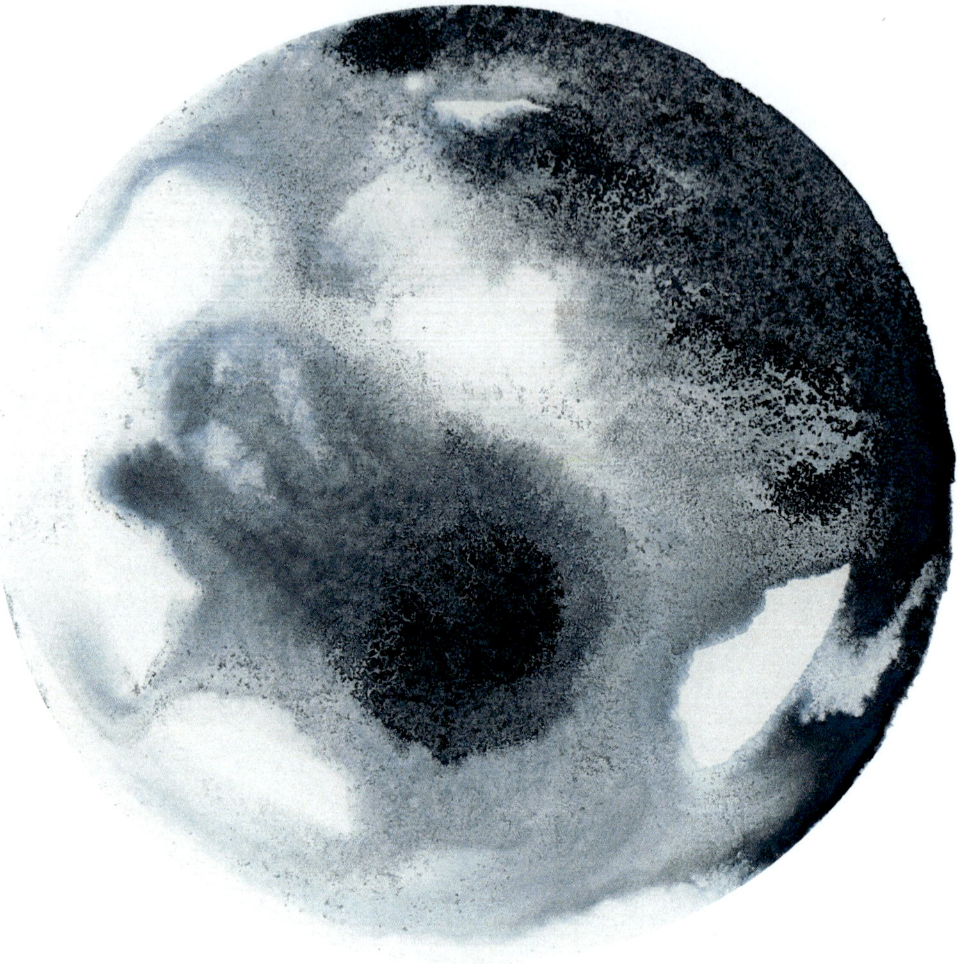

Os

76 osmium's toxic
Stains your skin with a blue-black tint
The most(?) dense of all the elements
Uncovers latent fingerprints

Group 8

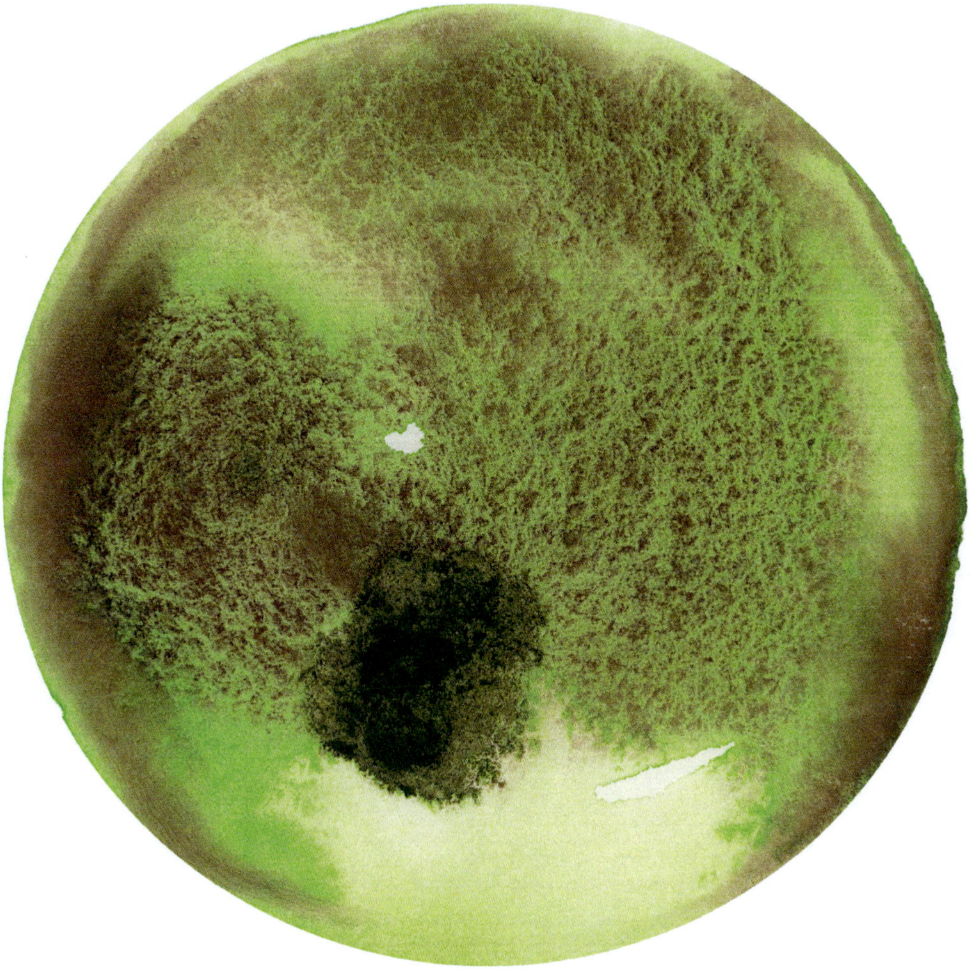

Hs

108 hassium's not found in nature
Shuffling cards is about its half life
We think at room temp it's probably solid
Made with atoms at the speed of light

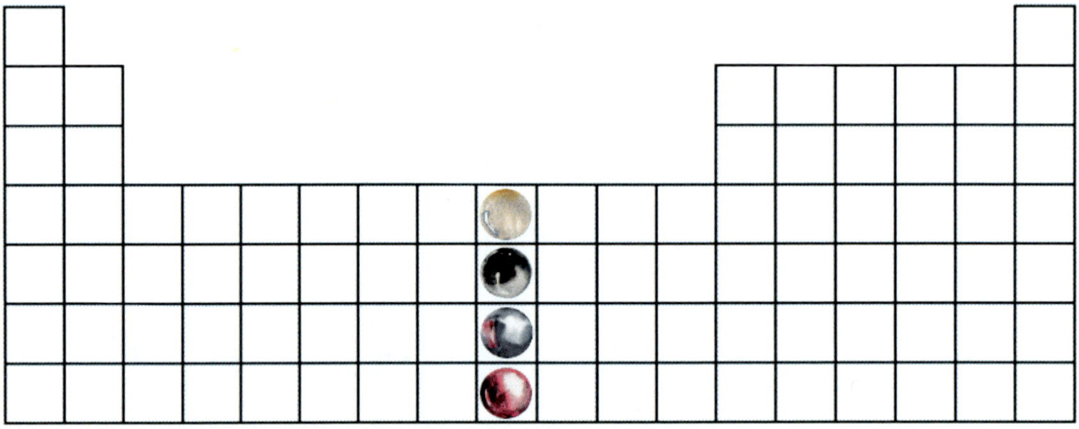

Group Nine

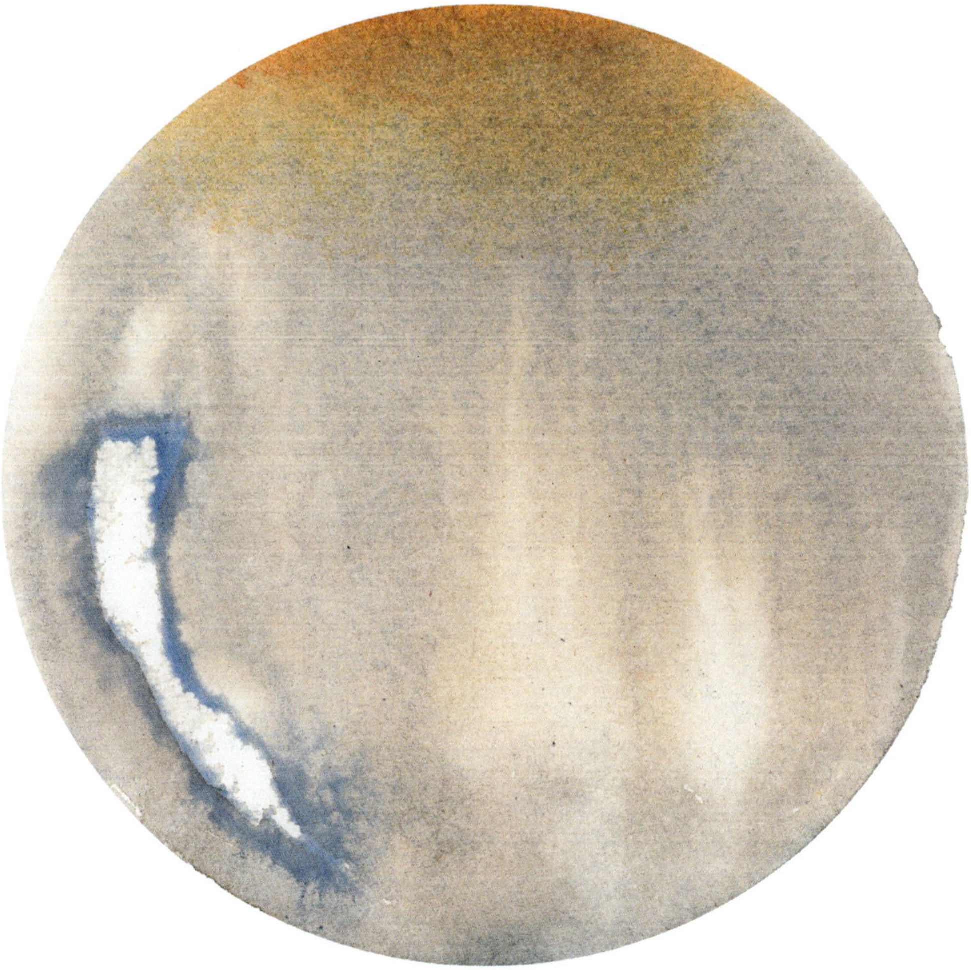

Co

27 is German for "goblin"
Kills cancer, shades high-end tattoos
Invisible ink once written in cobalt
In phones, drills, magnets, and blues

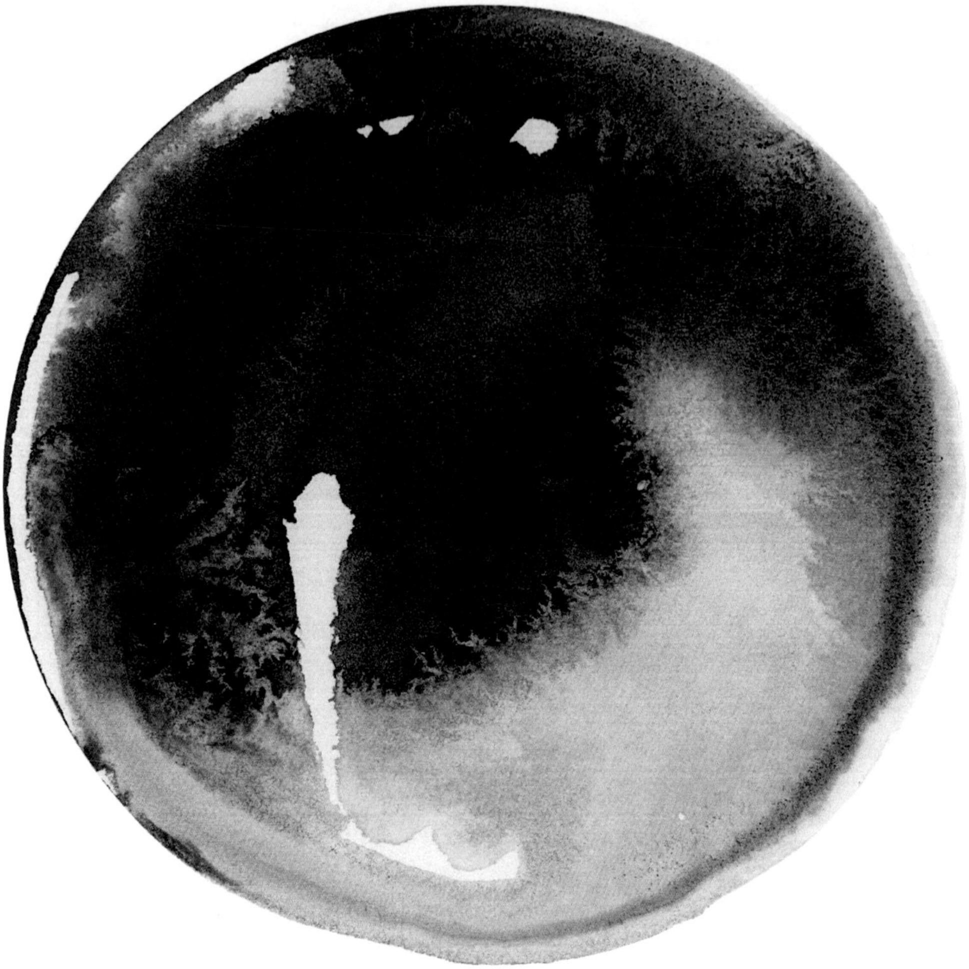

Rh

Rhodium is number 45
It knocks pretty rings out of budget
It's why your converter gets stolen
It's rare and acid can't touch it

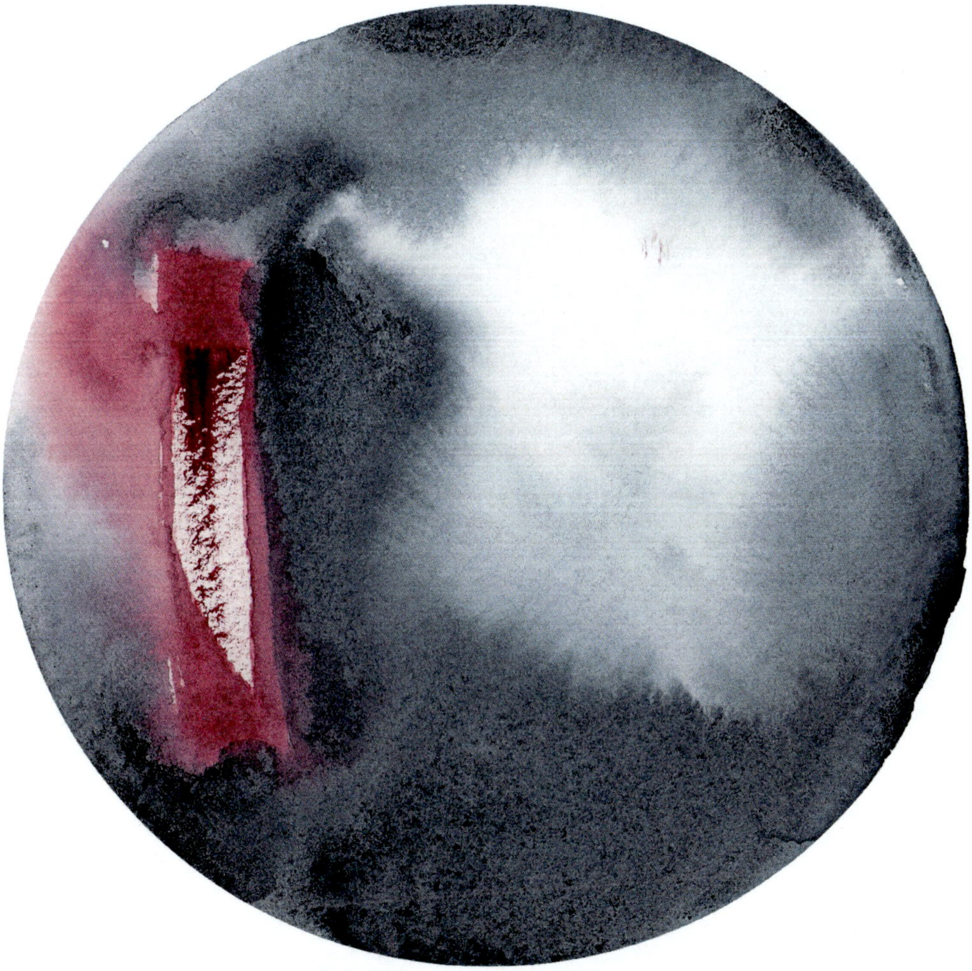

Ir

Iridium 77

It’s shimmery, tough, rare, and hard

Found in the K-T Boundary

In spark plugs for luxury cars

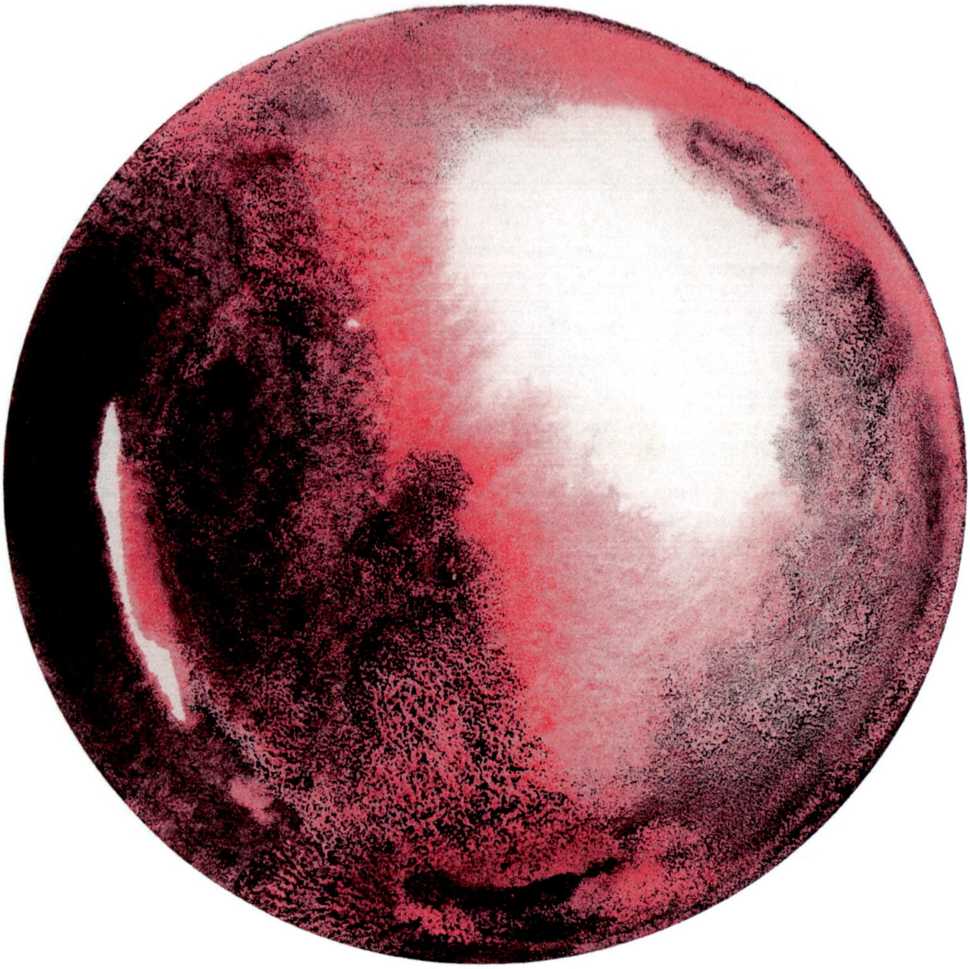

Mt

Artificial element number 109

Most of its facts are theorized

First solo woman honored with naming

"Meitnerium" makes up for a Nobel Prize?

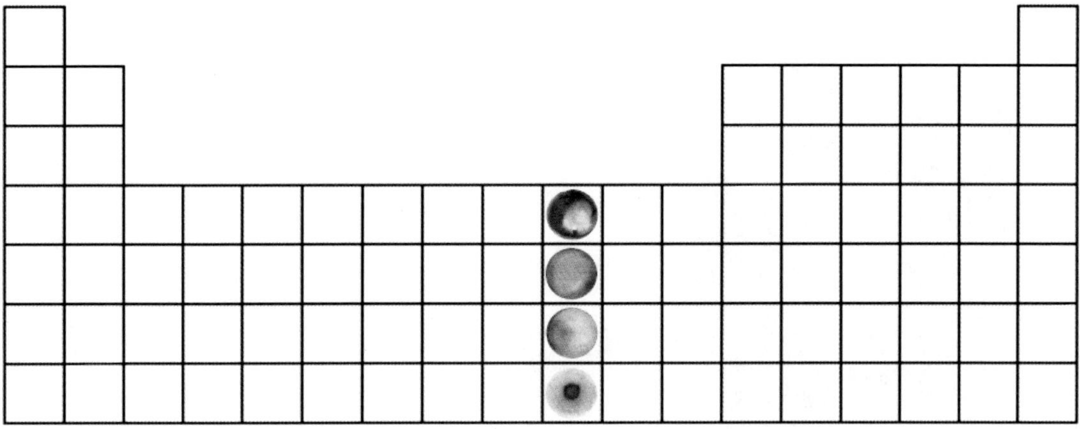

Group Ten

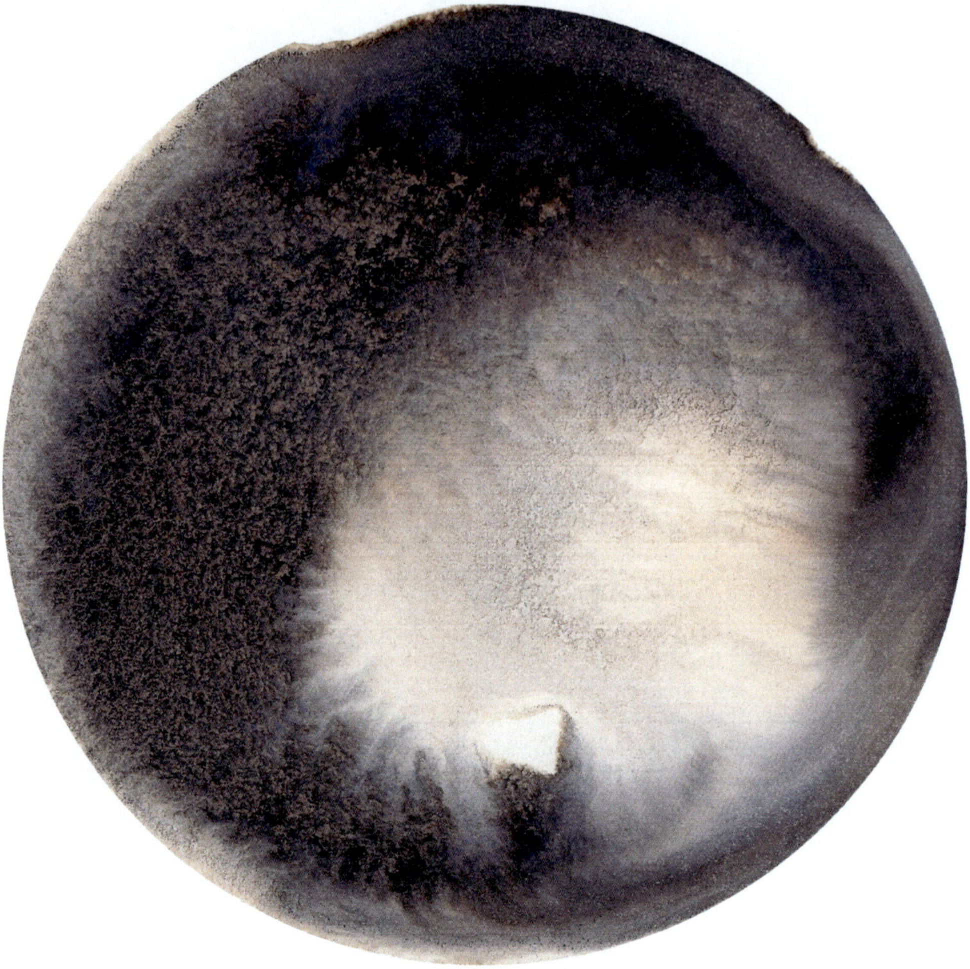

Ni

Number 28: “the devil’s copper”

Your nickel’s a quarter nickel

70% used in stainless steel

Found in cars and firms up your pickle

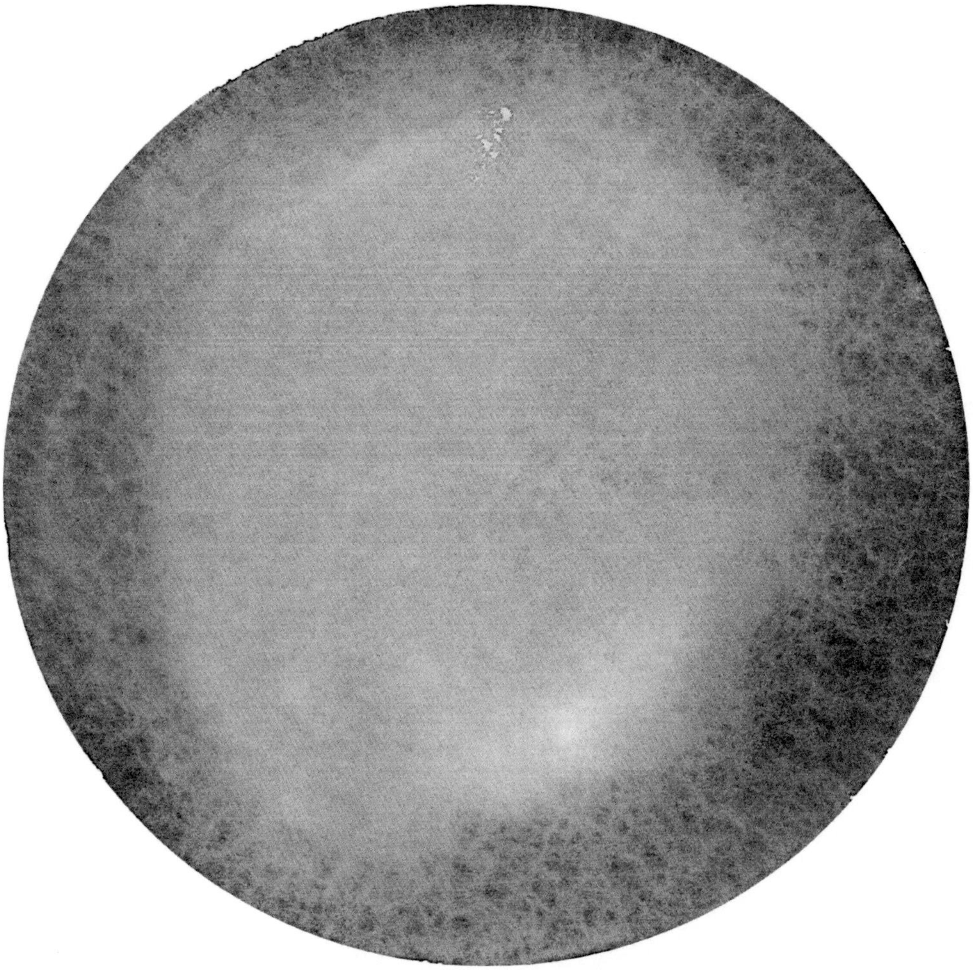

Pd

Palladium burns without a flame

46 controls cars' emissions

Atomic-level hydrogen sponge

No tarnish means pristine condition

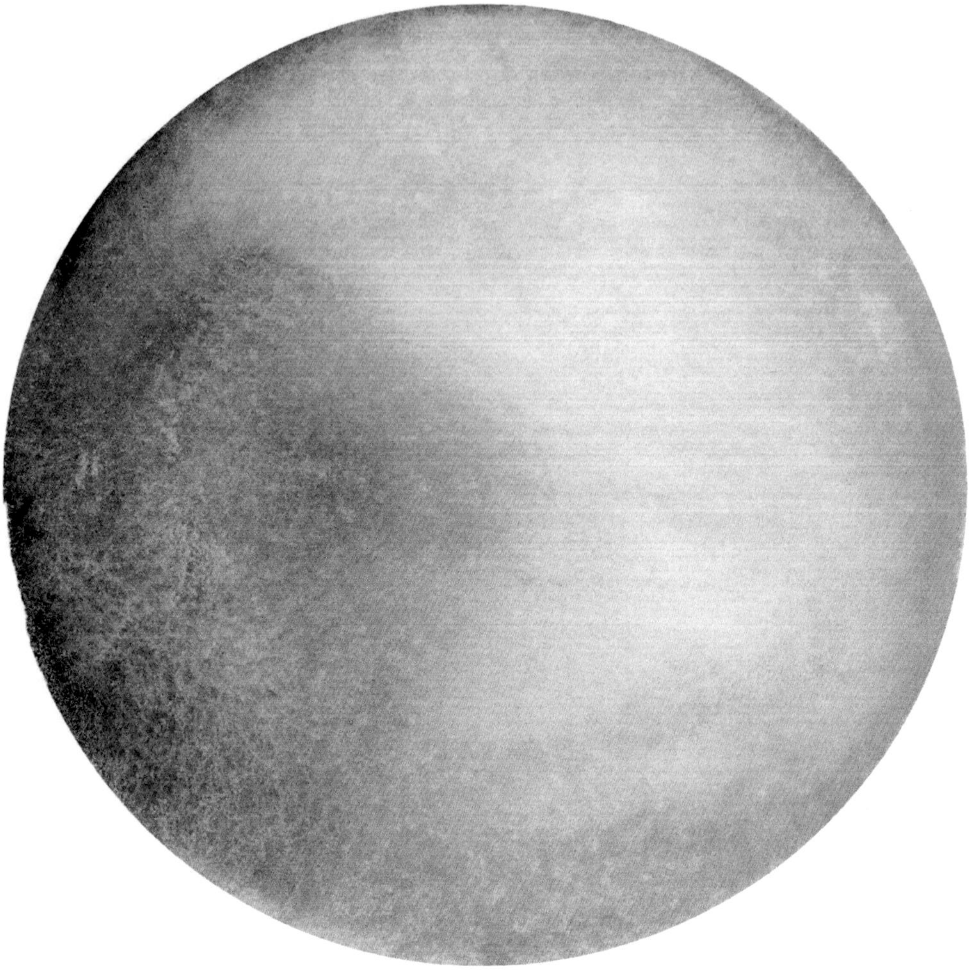

Pt

Defining mass and cancer's enemy
Platinum is number 78
More rare than gold, makes beautiful jewelry
Helps reduce your car's emission rate

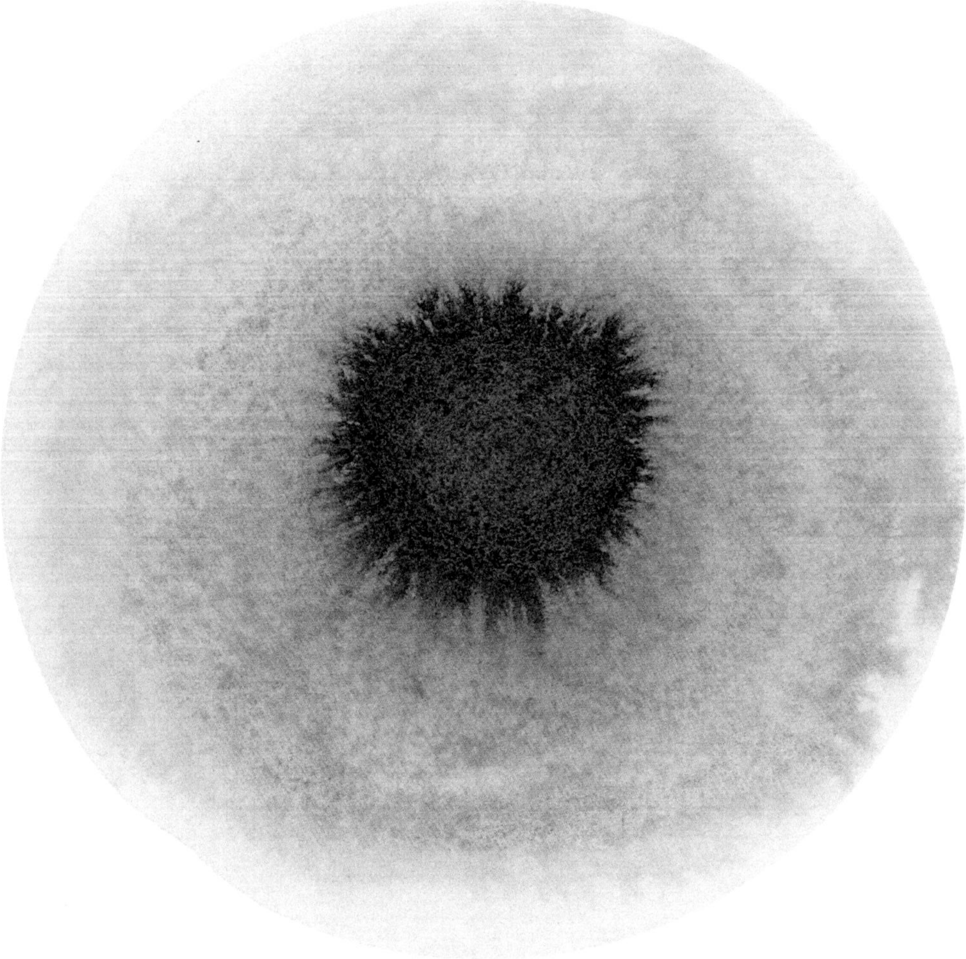

Ds

110 artificial darmstadtium
Only a few atoms produced
They sift through decay to ensure it was there
Everything else is deduced

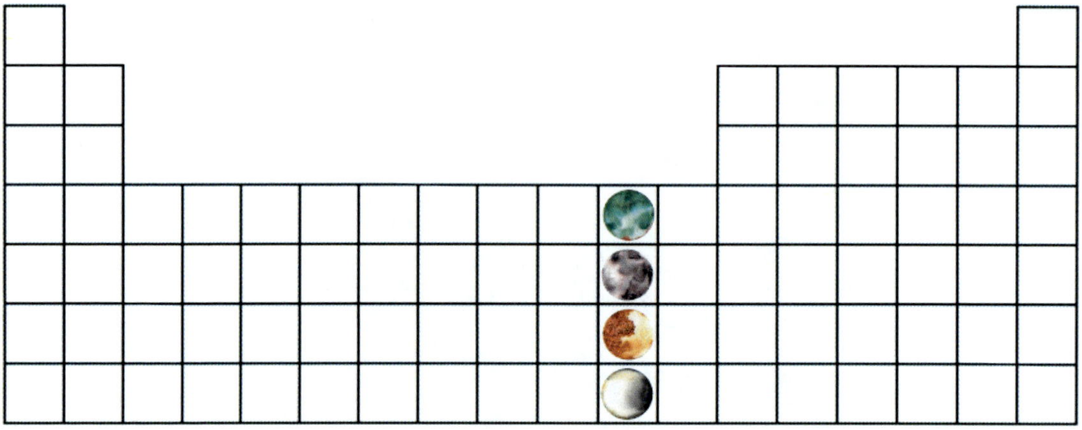

Group Eleven

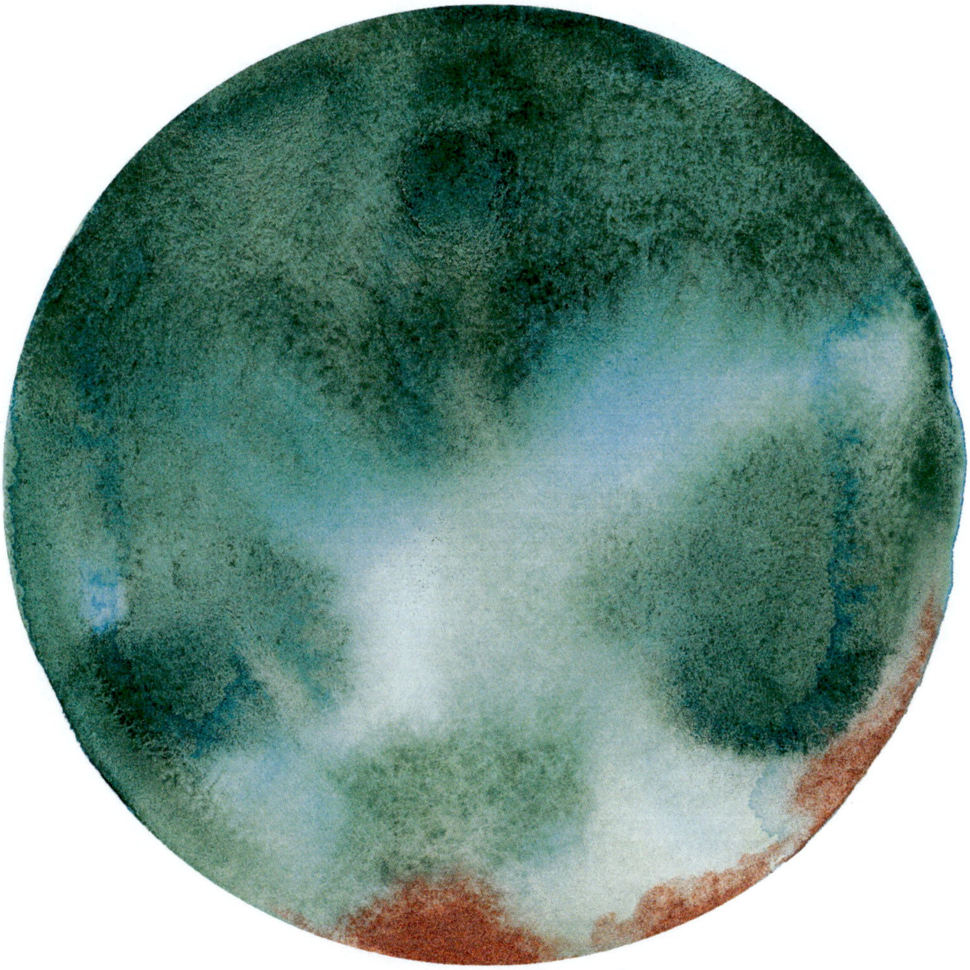

Cu

29: Most of your coins are copper

It’s a contact killing machine

In architecture, plumbing, and wiring

It's why Lady Liberty’s green

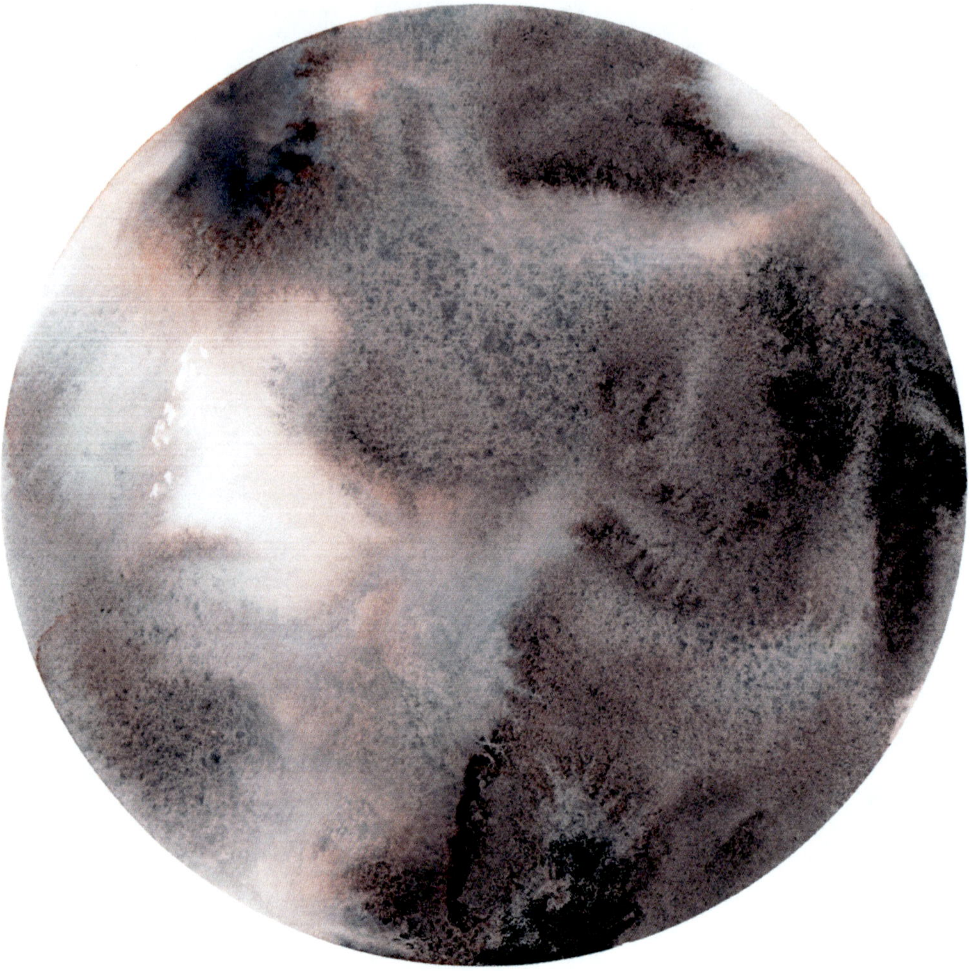

Ag

Silver takes gold in reflection
47’s first in conducting two
Its biggest flaw is its tarnish
Kills werewolves and germs (half true)

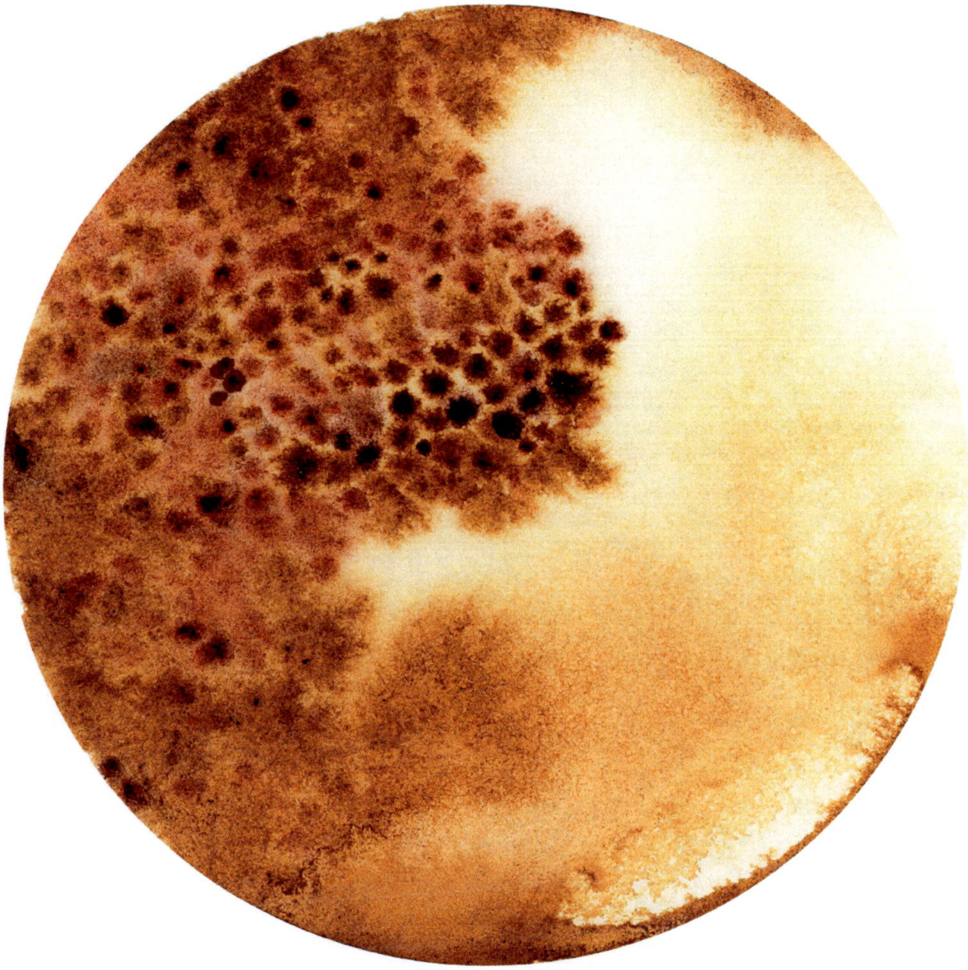

Au

Gold 79, forever shine

16 Psyche, jewelry, electronics

Malleable, edible, in humans

Was the basis for world economics

Rg

Lab-produced roentgenium 111

Named for the guy who discovered X-rays

Expected to be heavy metal

100 seconds before it decays

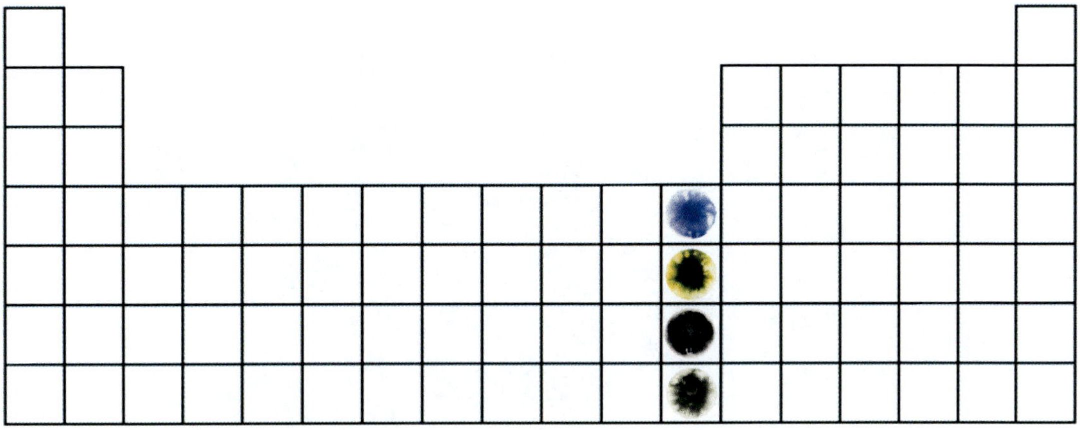

Group Twelve

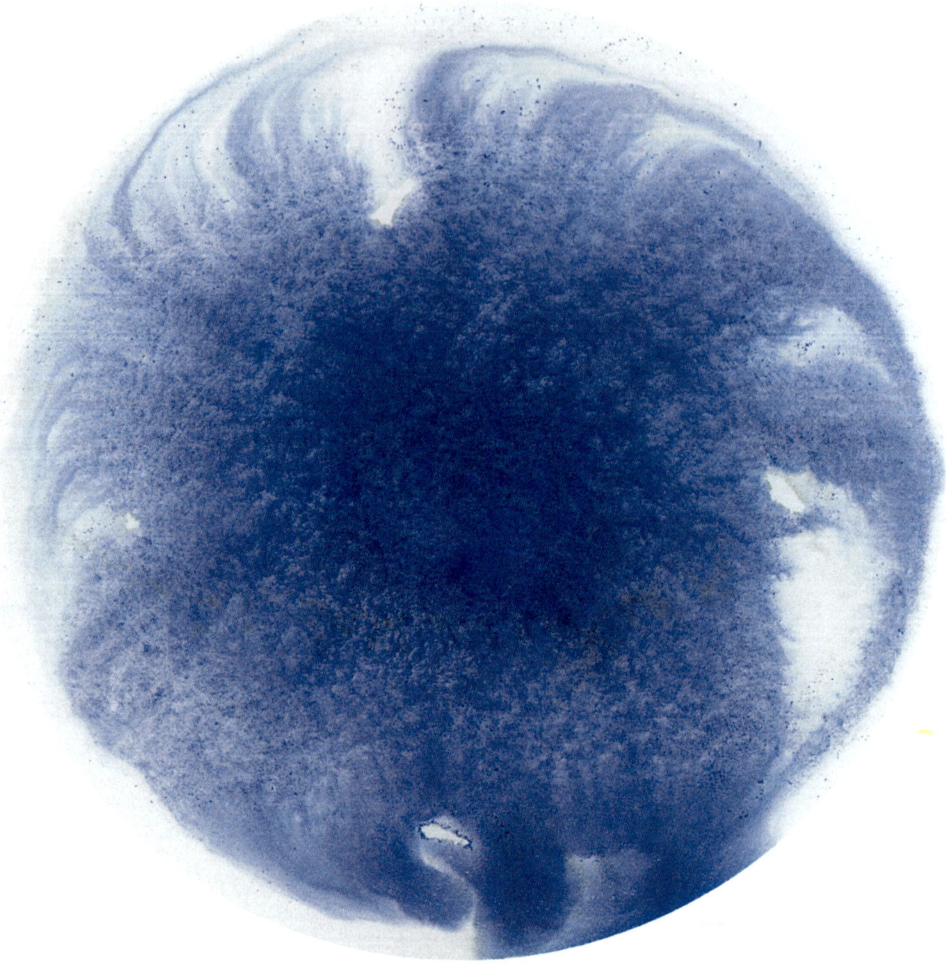

Zn

30 makes sacrificial anodes
It's zinc fingers, it's acid, it's base
Brass instruments, sunscreen, steel coating
It's crucial for scent, cents, and taste

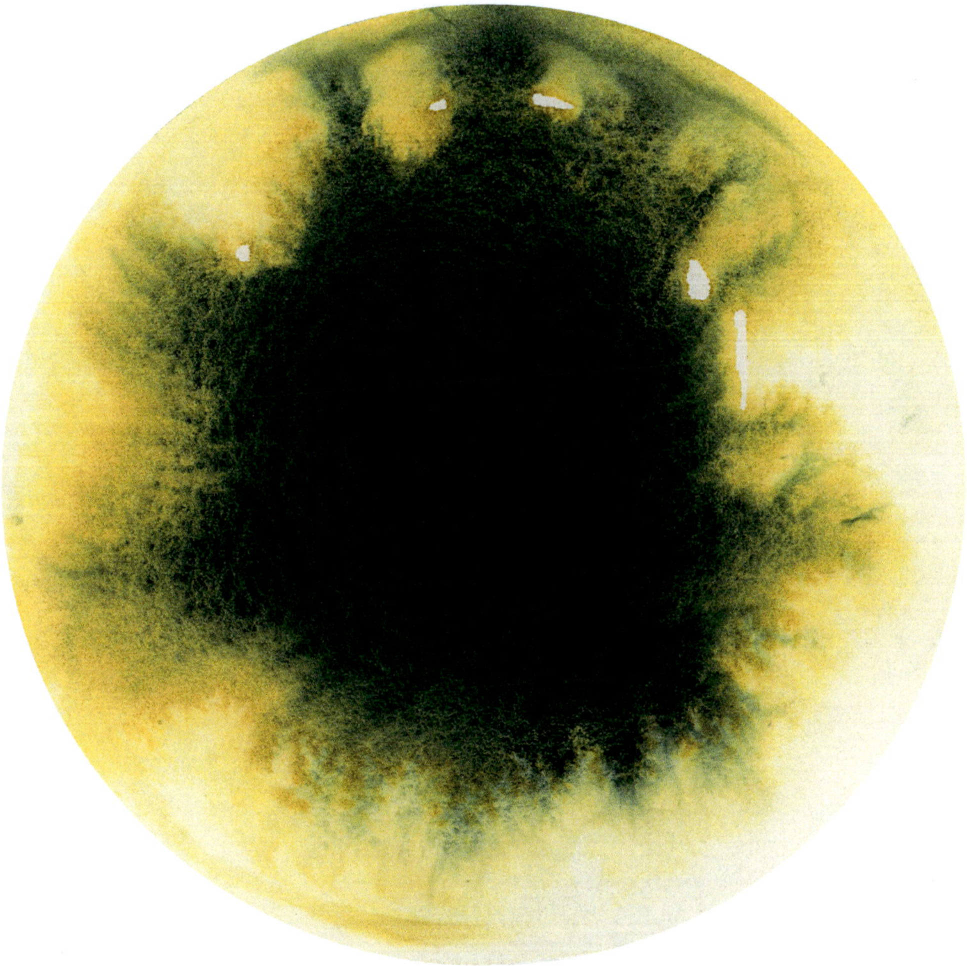

Cd

Number 48 is cadmium
Its colors made Monet swoon
It’s in airplane bolts and batteries
Keep it away from your food

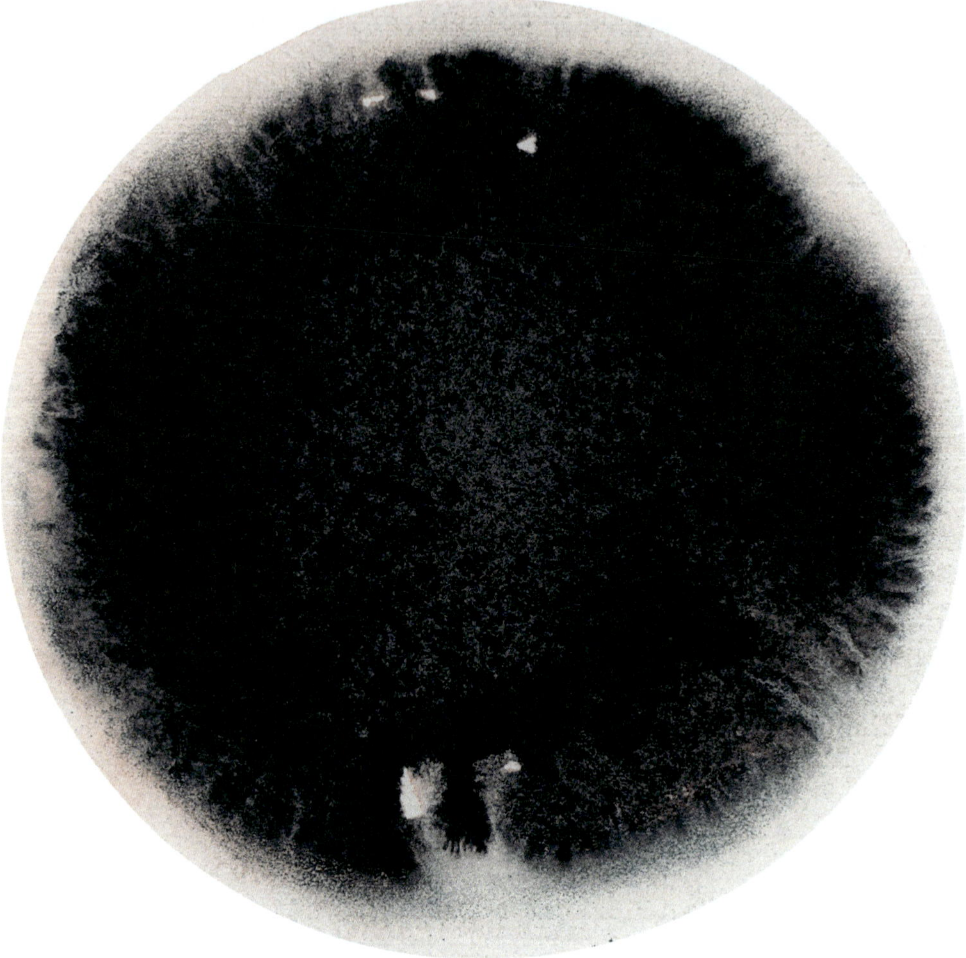

Hg

Group 12

Mercury 80; Quicksilver

Poisons and moves up the food chain

Dissolves gold and even floats lead

Dripping from cave walls in Spain

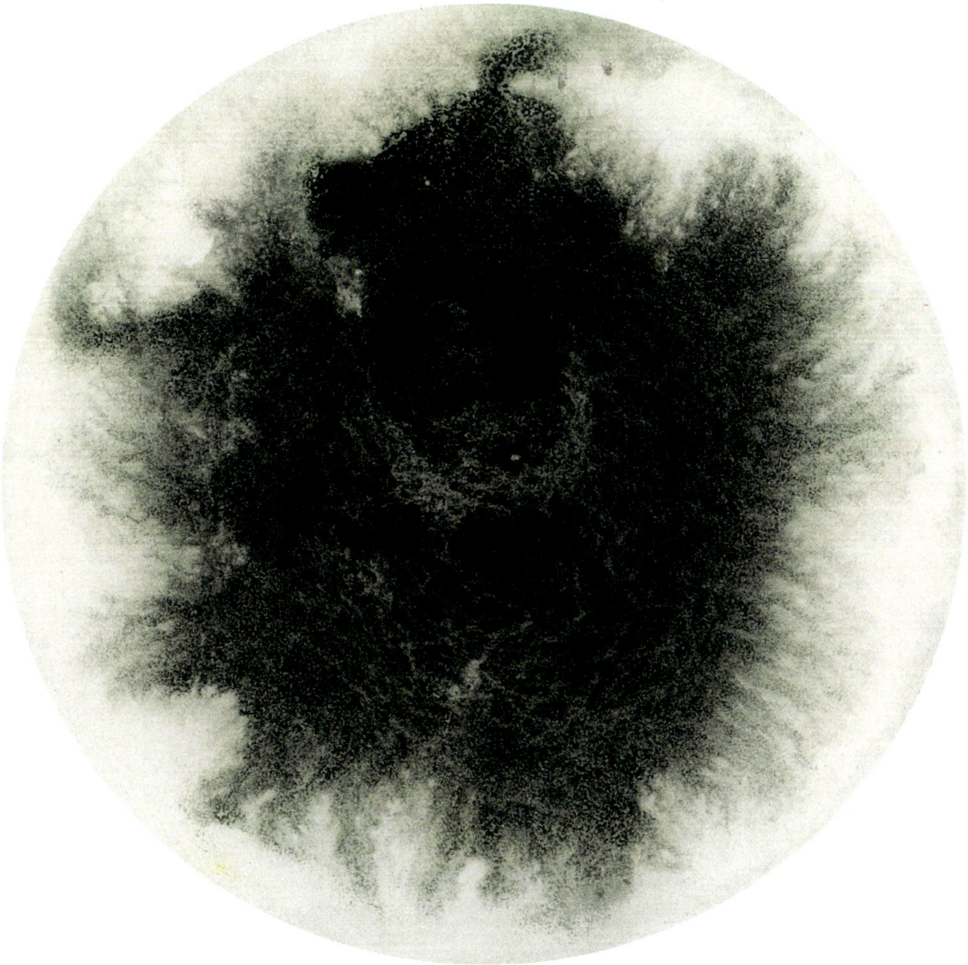

Cn

112 lab-made copernicium
The earth revolves around the sun
Half-life of an NBA shot clock
Made by pelting zinc ions

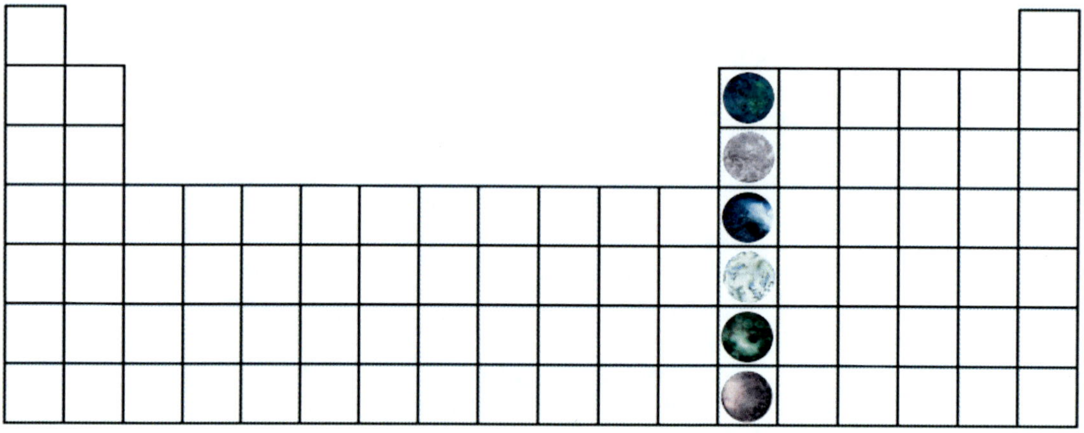

Group Thirteen

The Boron Group is electron hungry
Making cluster structures and dimers
Hard, soft, softer, more soft, softest, vapor
This family is made of outliers

Group 13

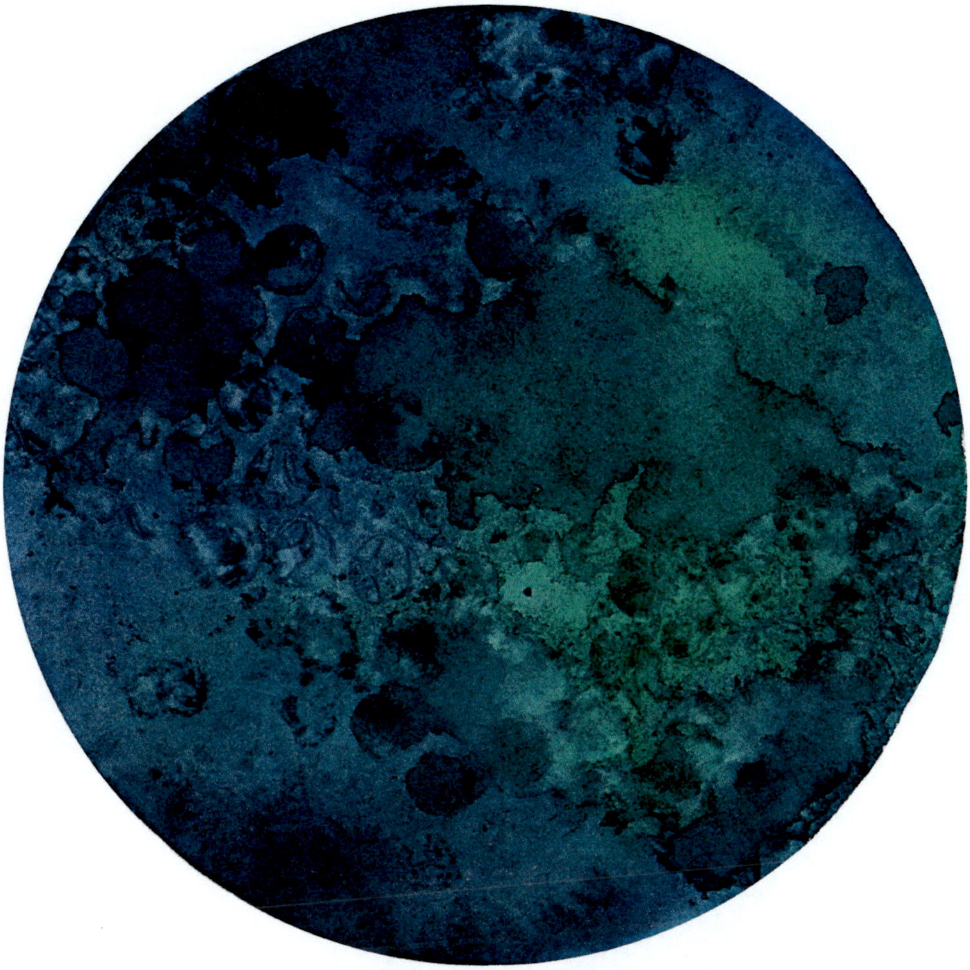

B

Boron 5 makes a IIb diamond
This metalloid makes fireworks green
Forms covalent bonds and Pyrex
Treats cancer and gets your clothes clean

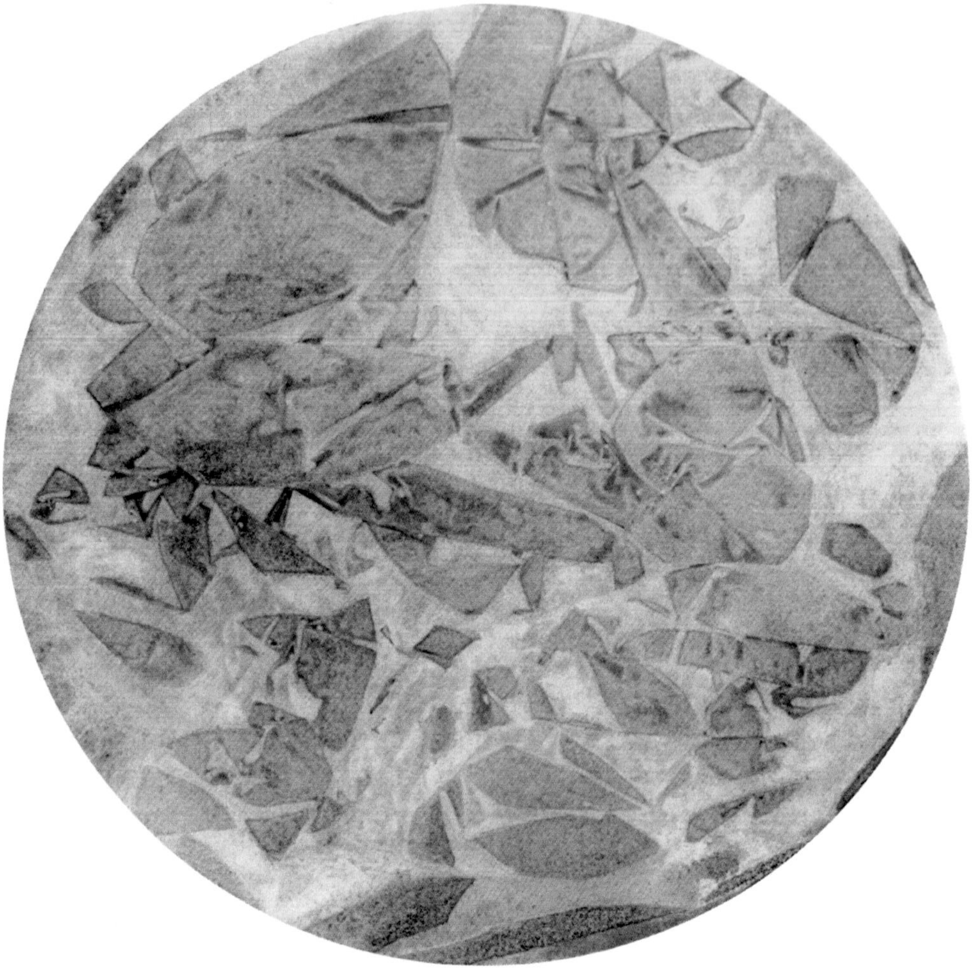

Al

Aluminum's light, strong, and won’t rust
Once more costly than gold (it’s 13)
In airplanes, rocket fuel, and foil
Really effective recycling

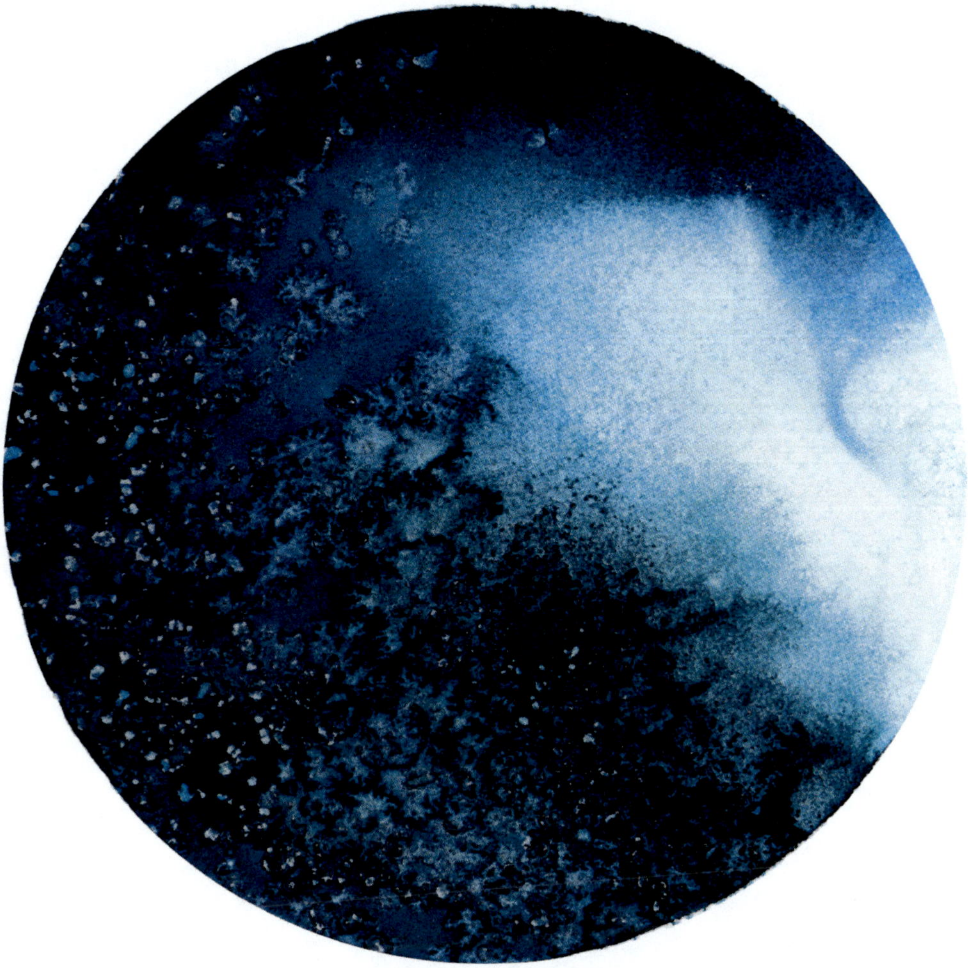

Ga

Element 31 is gallium

It'll melt if you grab a fistful

Expands 3% as a solid

Makes semiconductor crystals

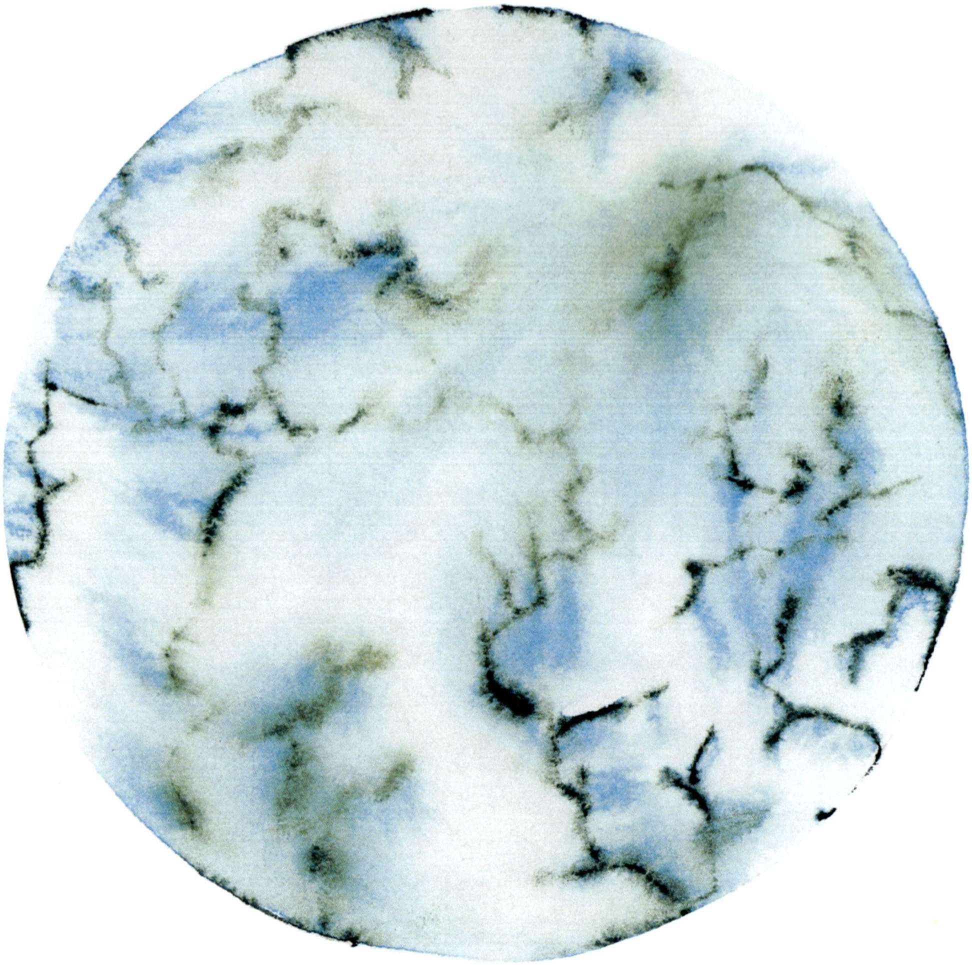

In

You'll hear a scream when you bend indium
49 will make glass wet
It conducts with no corrosion
ITO's getting harder to get

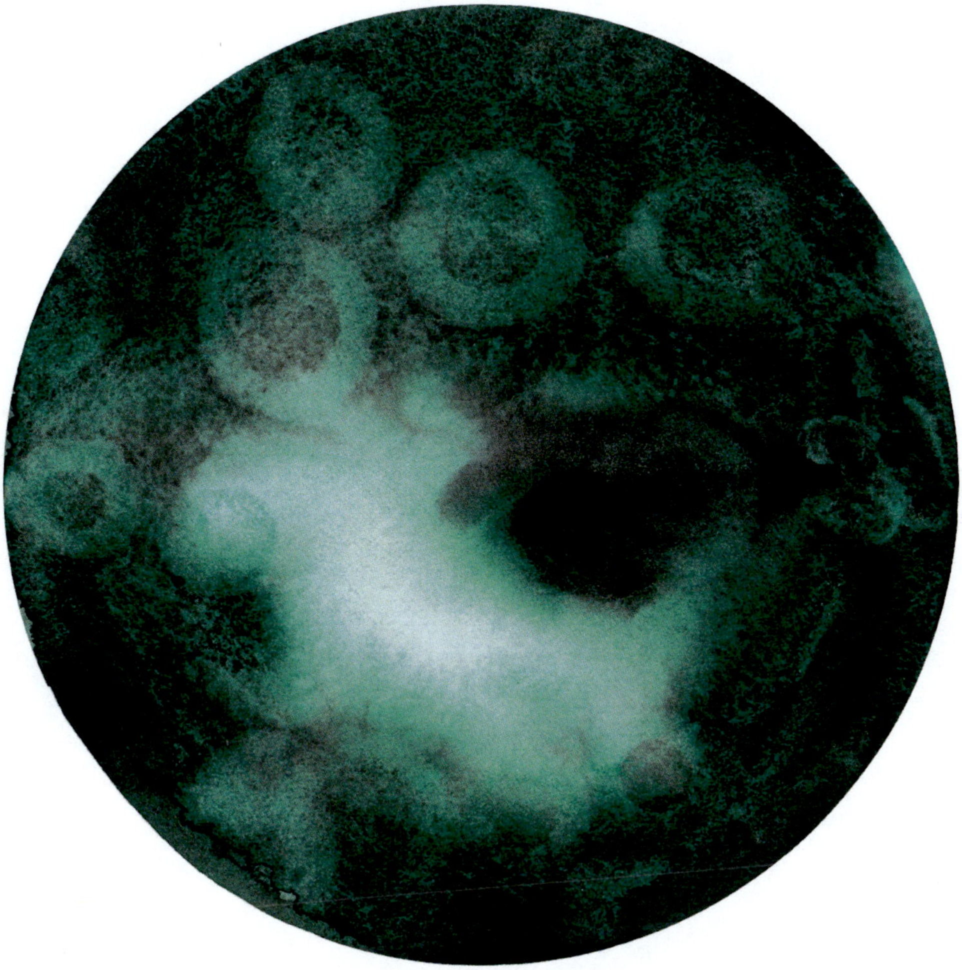

Tl

Element 81 is poison
Murder by thallium was once en vogue
Infrared optics and lenses
Heat it up for a soft green glow

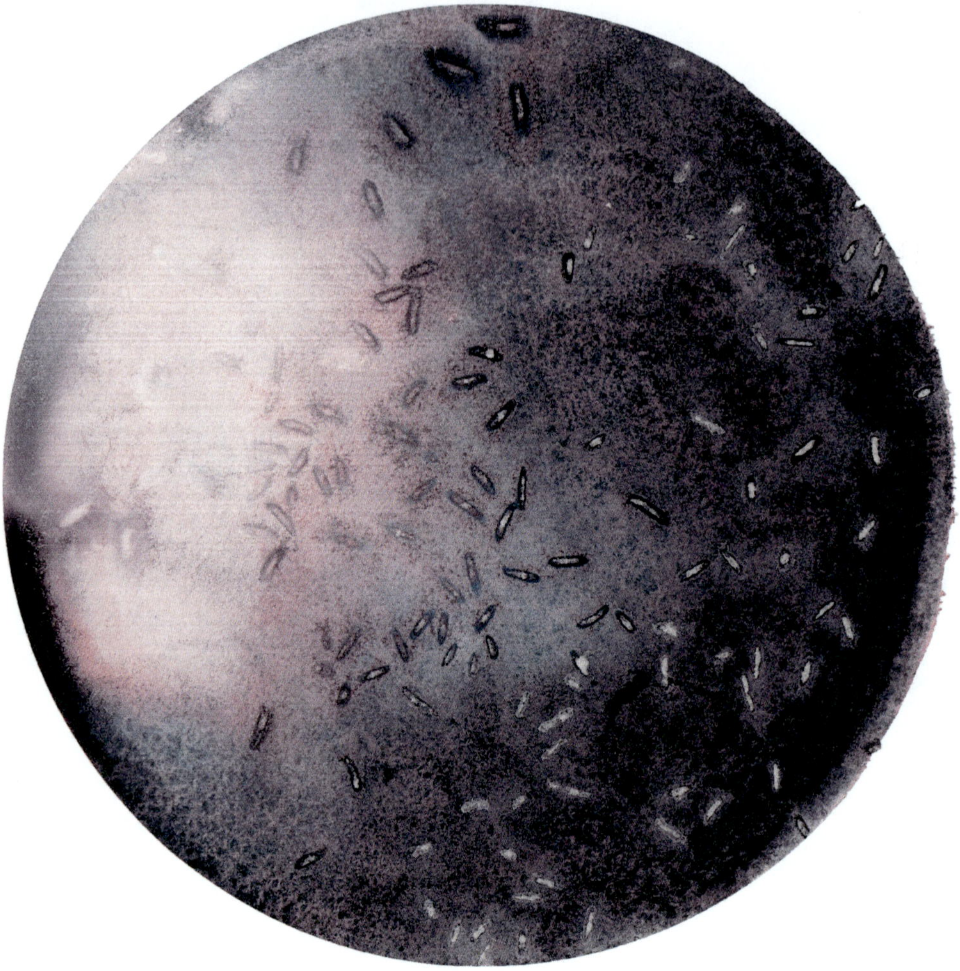

Nh

Nihonium number 113

Half life; 20 seconds or less

Created by bombarding atoms

Most that we “know” has been guessed

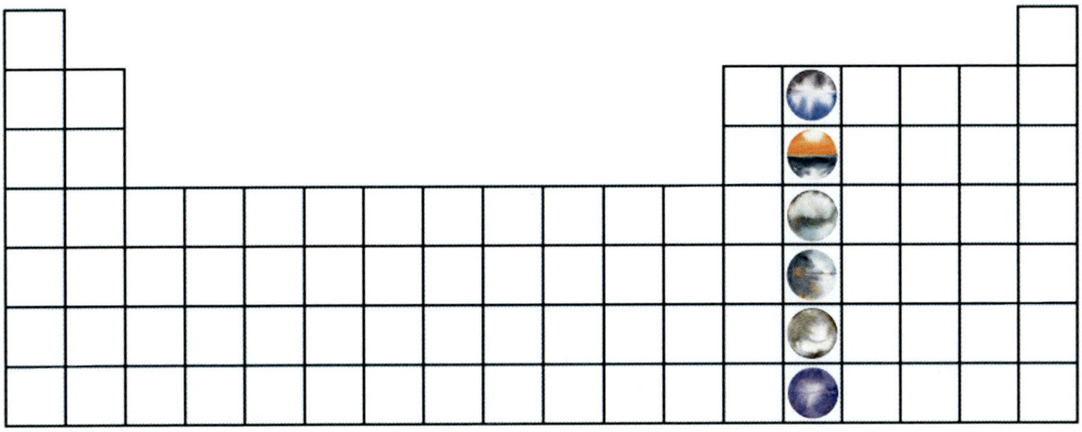

Group Fourteen

The Carbon Group is your ride or die

They all have 4 valence electrons

Single, double, or triple;

Adaptable, flexible bonds

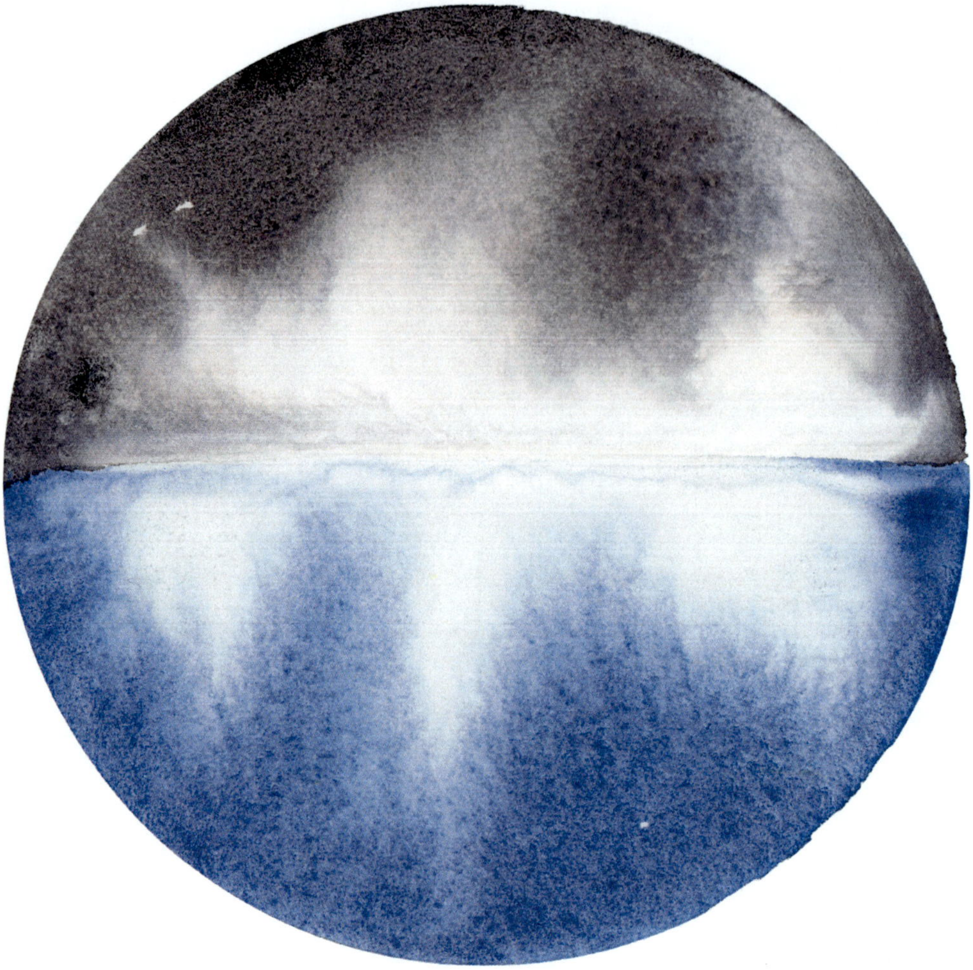

C

The anchor of life - number 666

Carbon bonds with itself to make chains

Forms diamonds, and humans, and dry ice, and farts

End its cycle and watch climate change

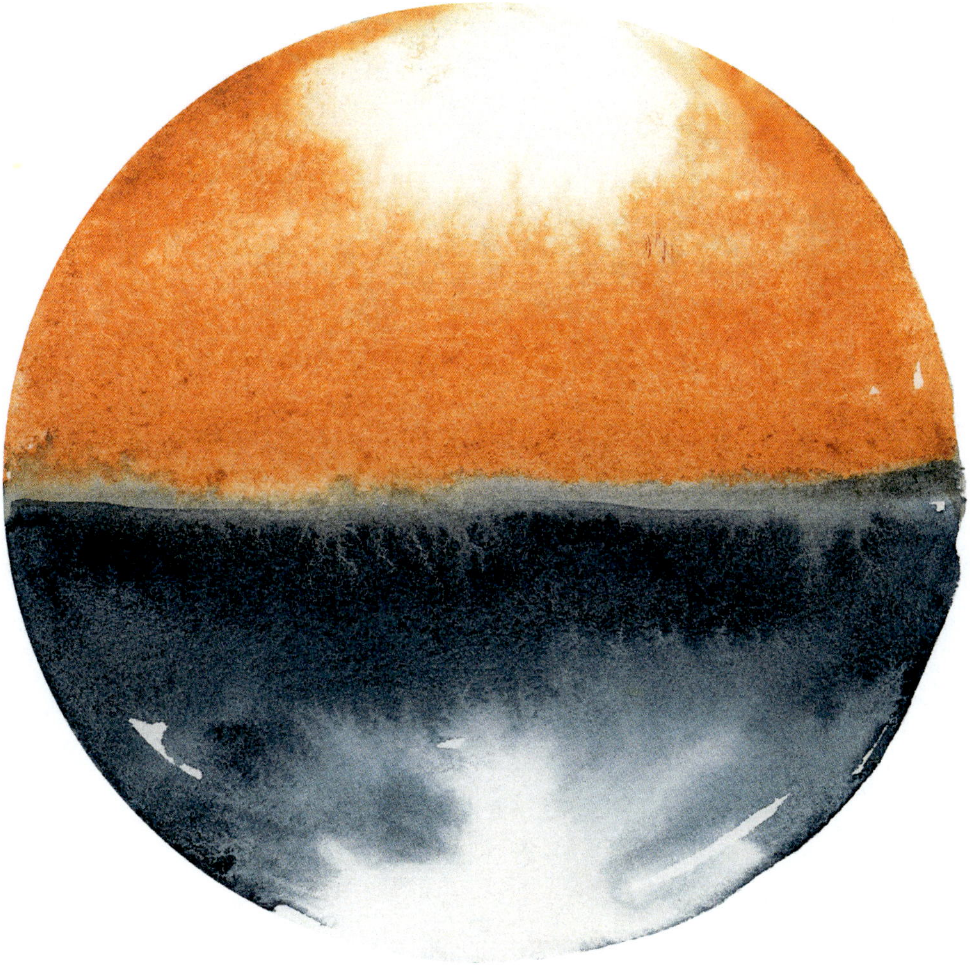

Si

Number 14 drives the digital age
Silicon Valley was named after it
Desiccant packets and breast implants
A grain of sand to the microchip

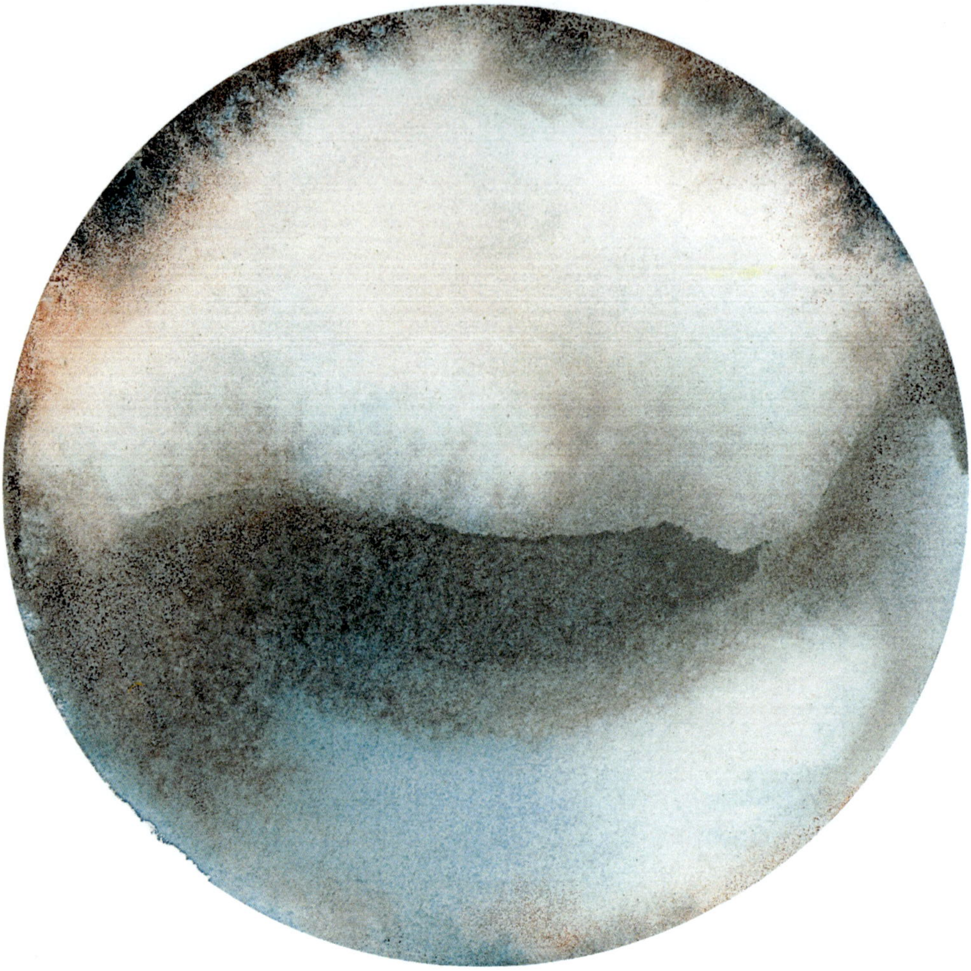

Ge

Germanium; number 32
Was a blank spot on the table
The queen of early transistors
Infrared and optic cable

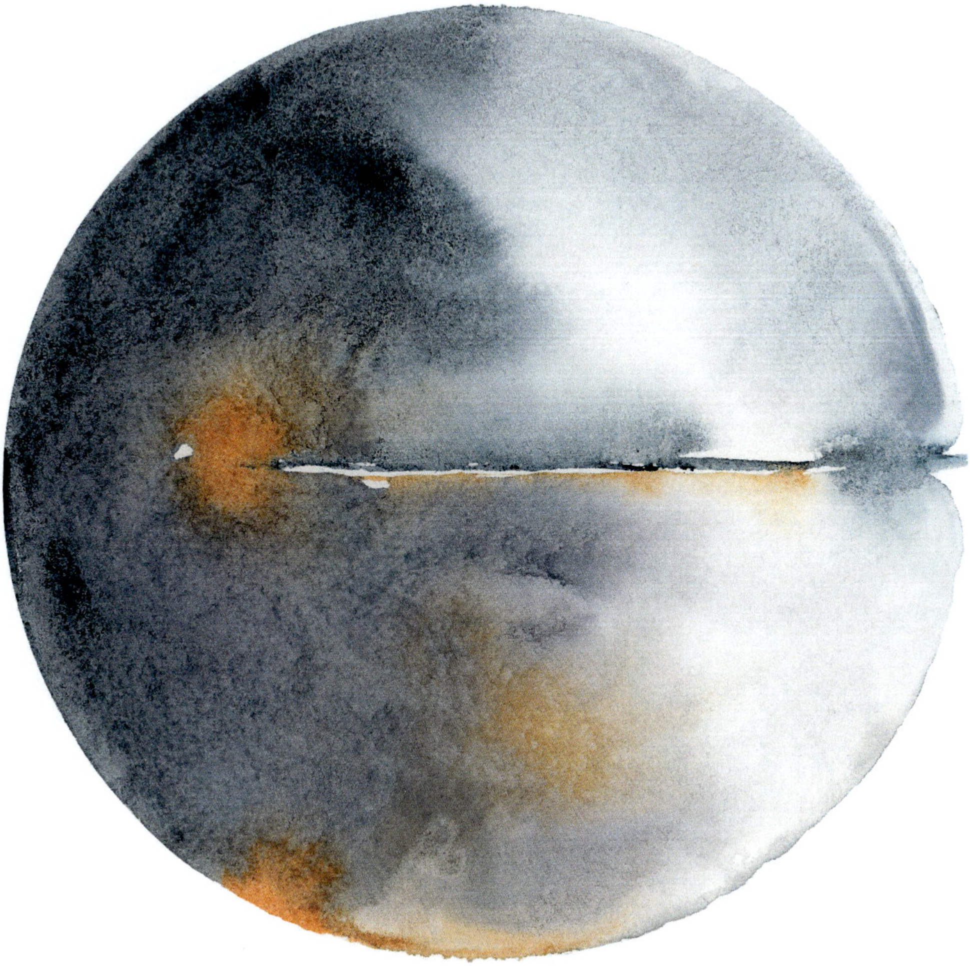

Sn

Tin 50, the most stable isotopes
It cries when it gets twinned
Once used as valuable currency
Can, foil, Man - aren’t tin

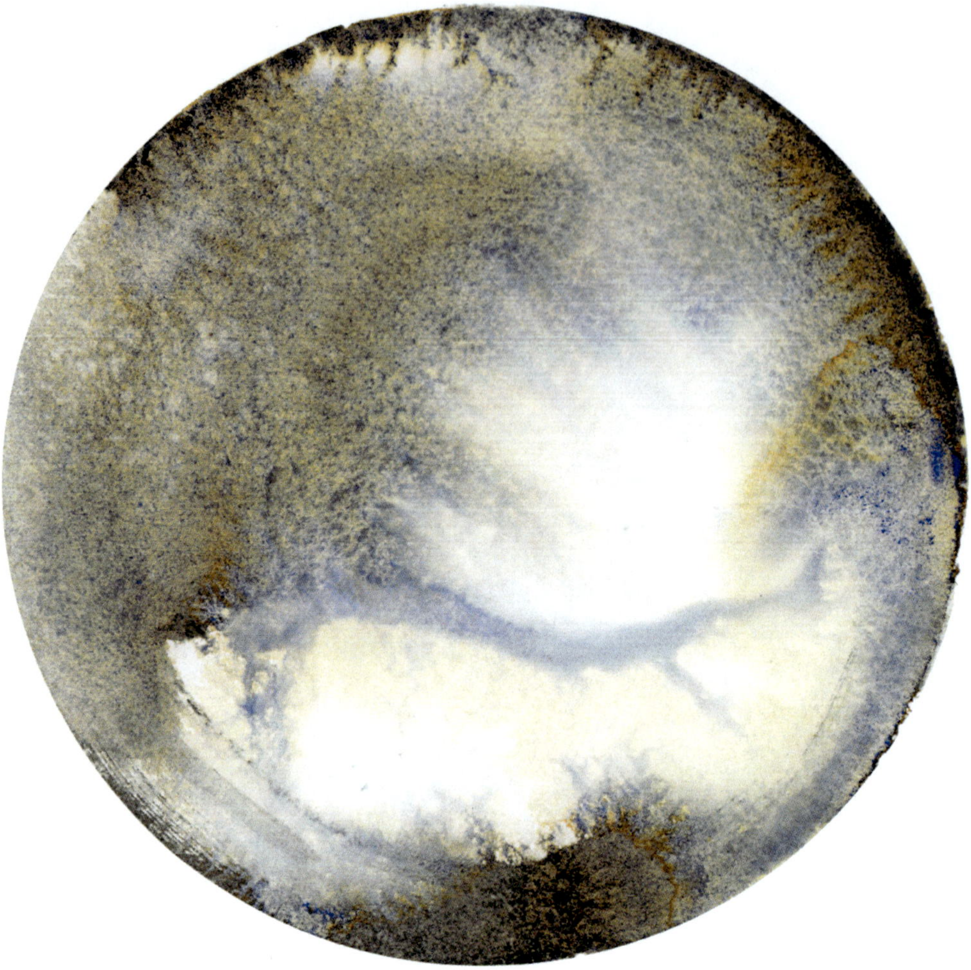

Pb

Group 14

Lead is element 82
Makes a great radiation shield
Super toxic (especially in kids)
That's why leaded gas was repealed

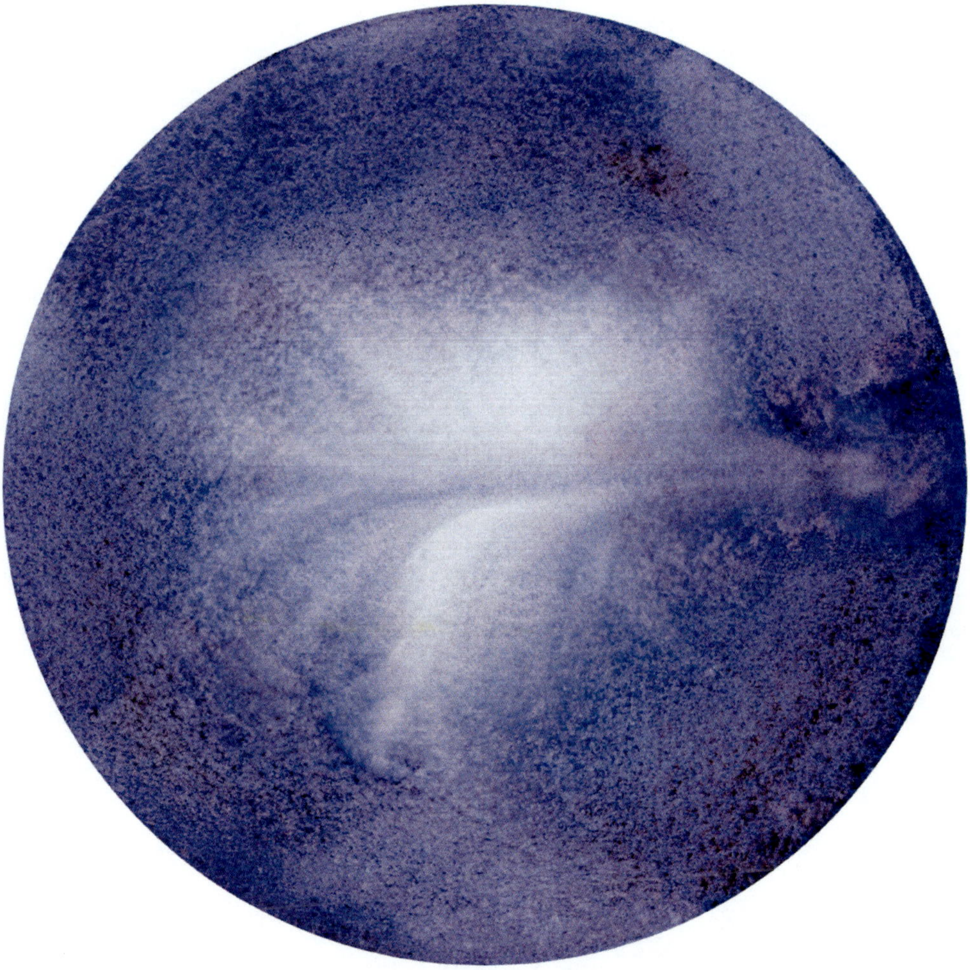

Fl

Flerovium is mostly predicted
We know for sure it's 114
5th heaviest element ever
Redefined what atoms can be

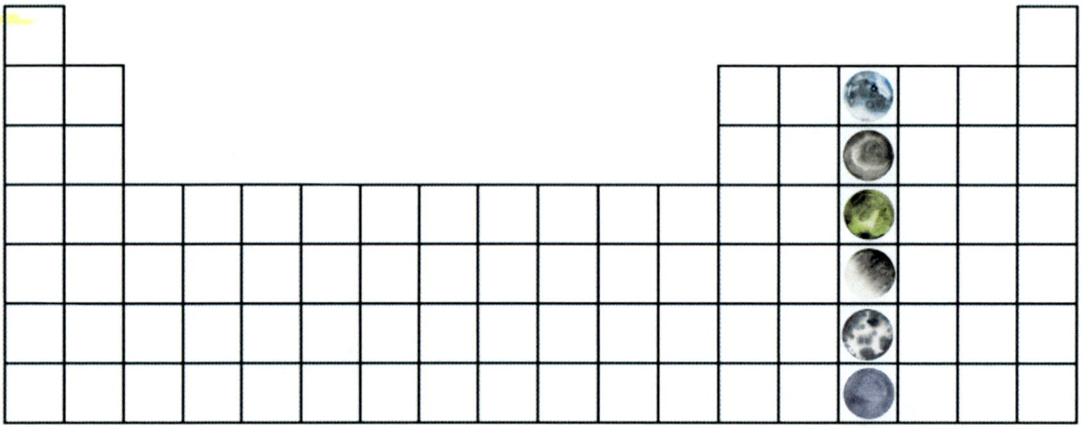

Group Fifteen

The Nitrogen Group forms MH3

Loves to take on 3 electrons

It's a real mishmash of matter

It's eager for covalent bonds

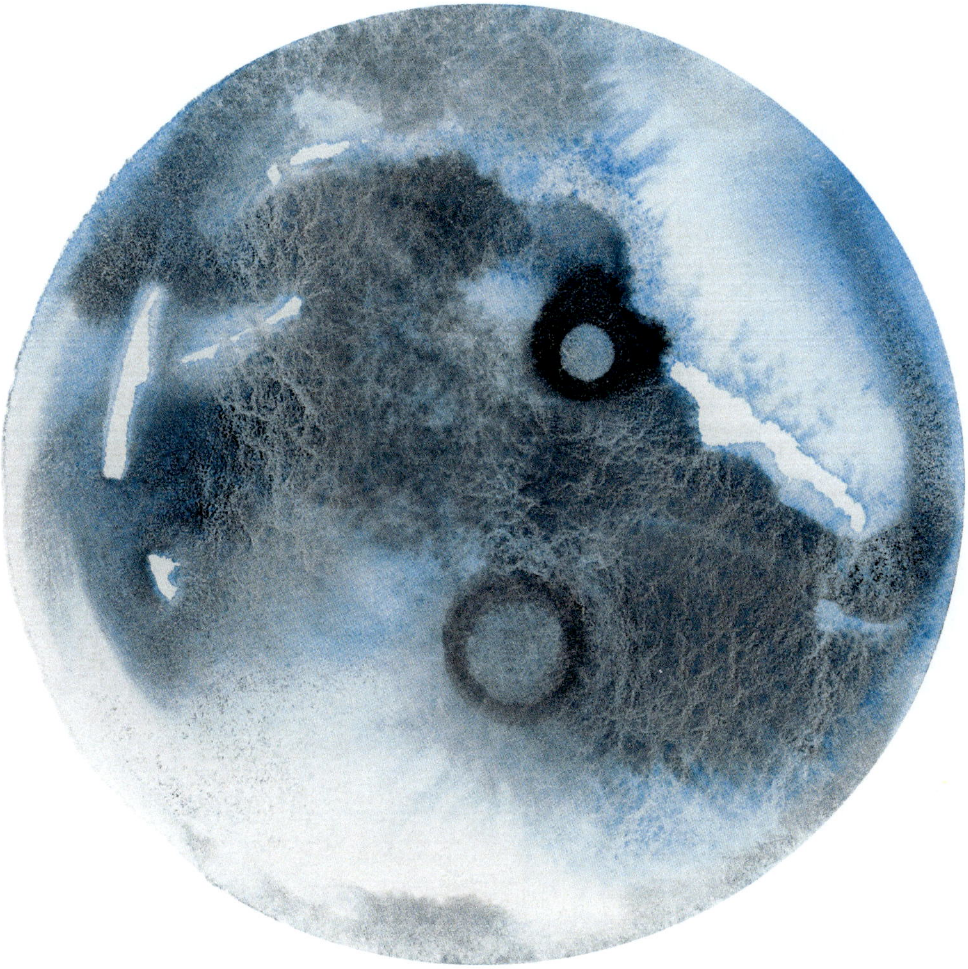

N

Element 7 is nitrogen

78% of the air you breathe

Triple bonds means it’s safe (but explosive!)

Feeds ⅓ of the world, makes ice cream

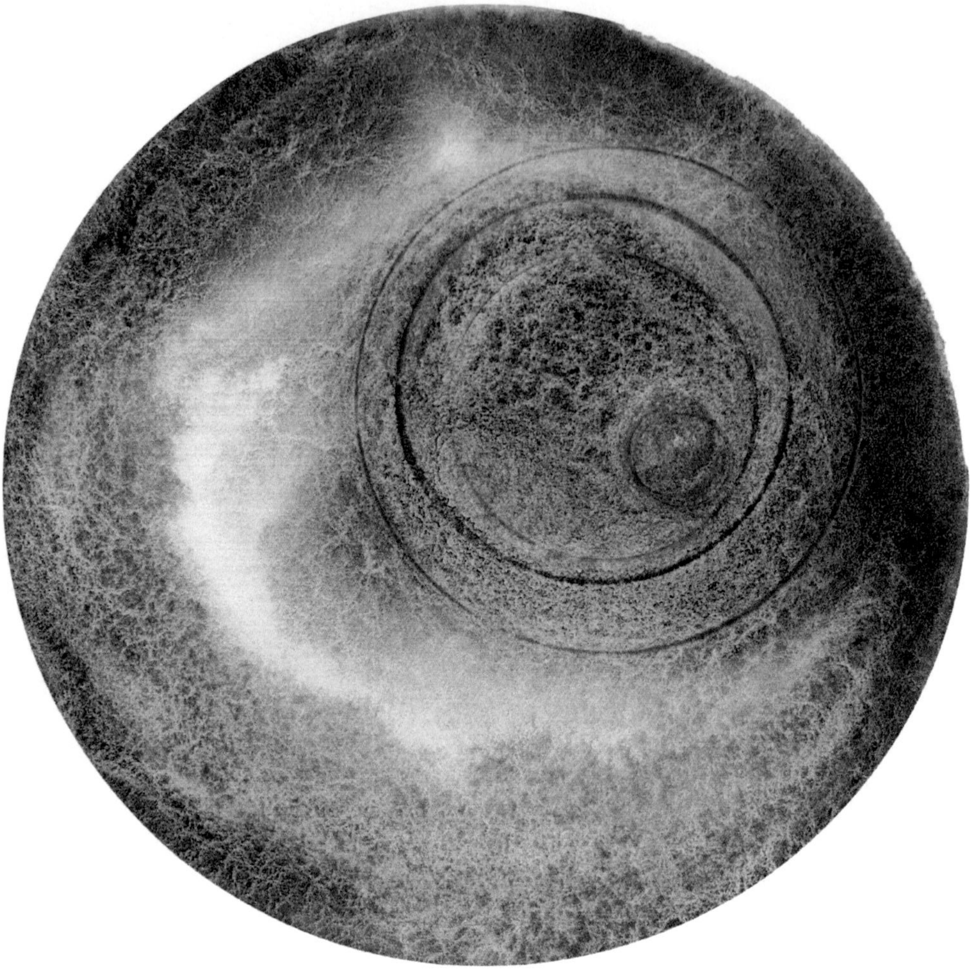

P

Vital to life and close to pure evil

Phosphorus is element 15

Combusts in air and burns underwater

Discovered while working with pee

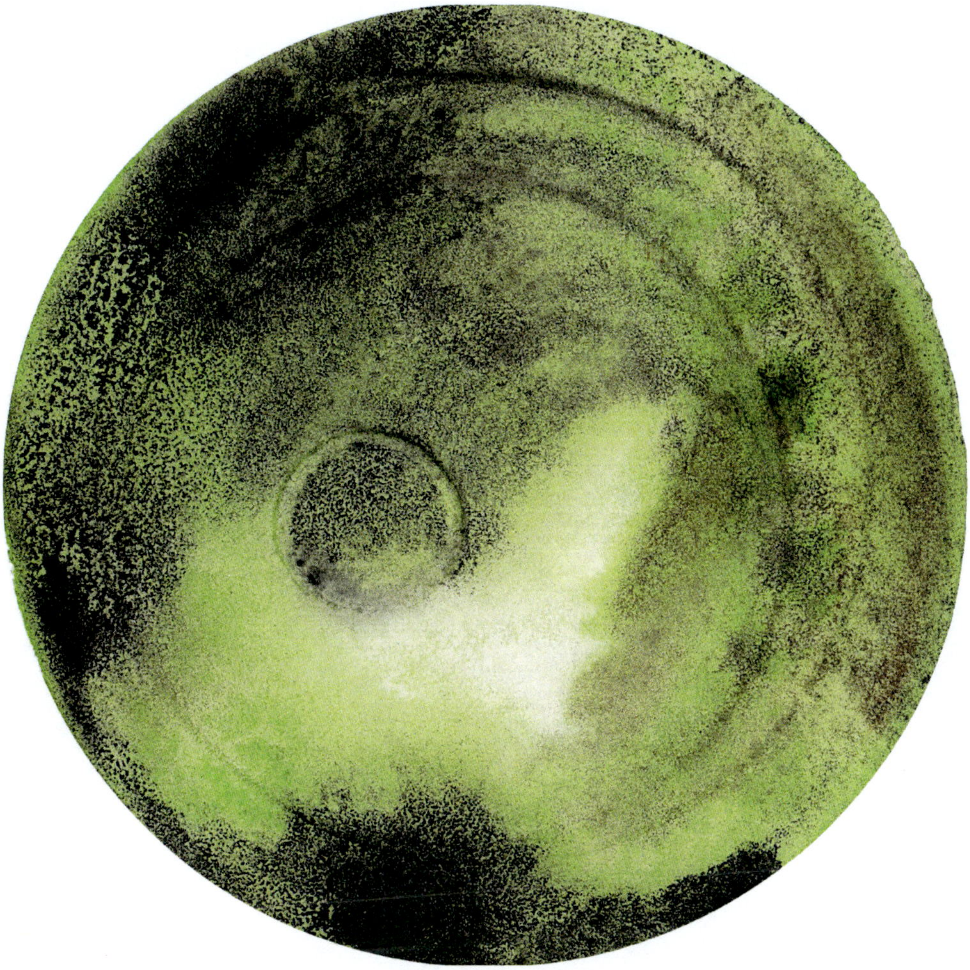

As

Death by wallpaper, the Poison of Kings

Arsenic is number 33

Treats cancer and contaminates water

It's in optics and chicken feed

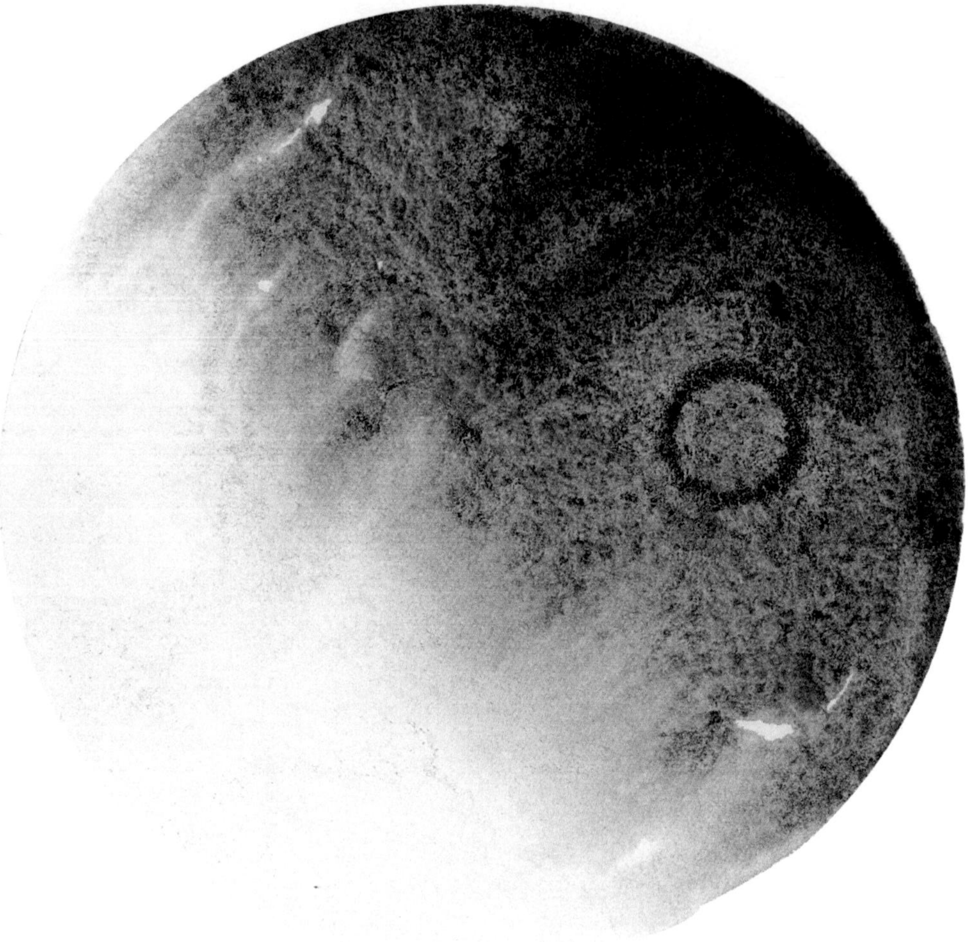

Sb

Antimony; Element 51

When it cools, it sings and expands

Bullets, eyeliner, and moveable type

Was it Napoleon's end?

Group 15

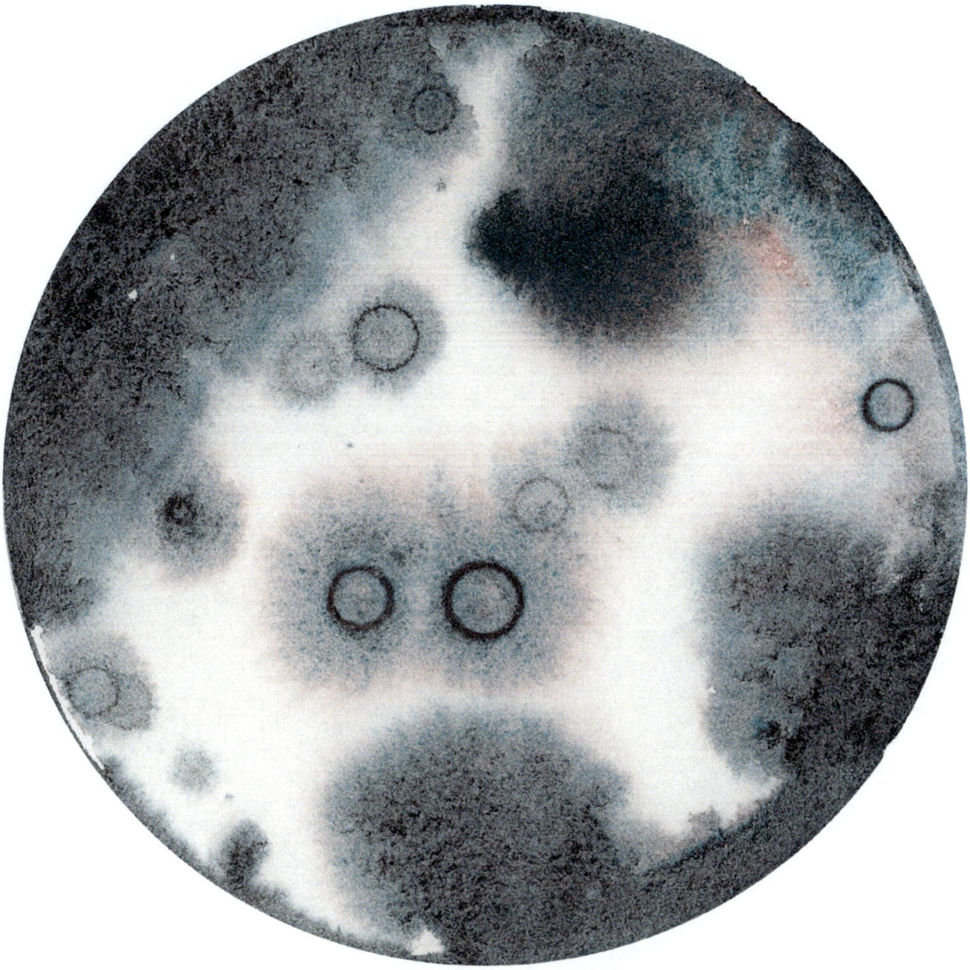

Bi

Bismuth is element 83

Fire alarms and rainbow hopper crystals

Heaviest (mostly) stable element

Turns poop black when you drink Pepto-Bismol

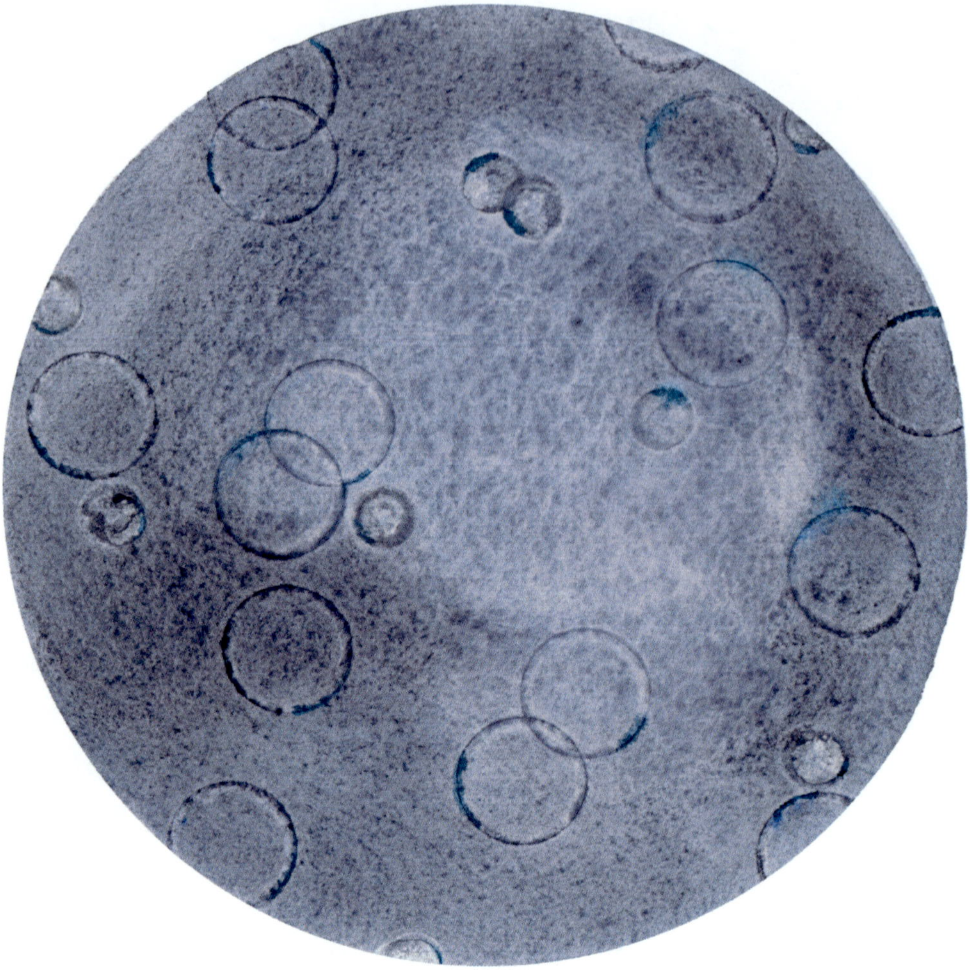

Mc

Moscovium is 115

Highly unstable condition

First synthesized in Russia

Everything else is prediction

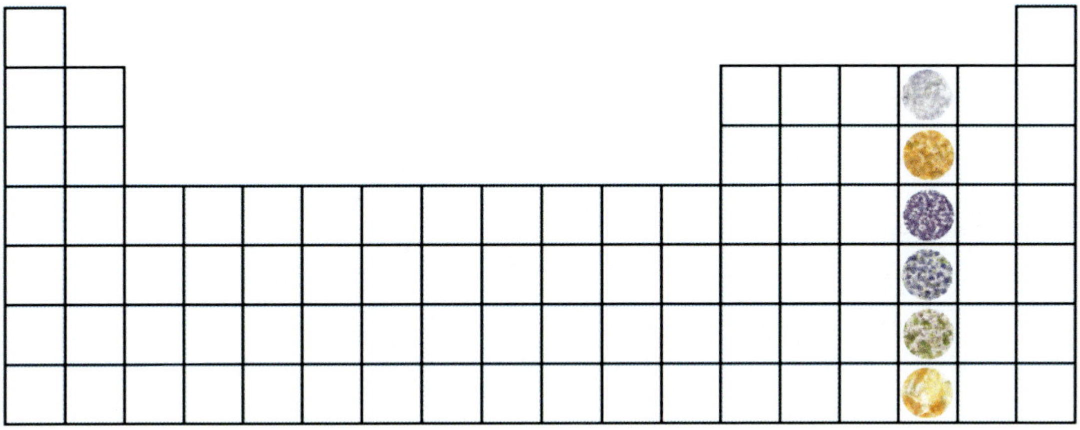

Group Sixteen

The chalcogens are allotropic

And 2 electrons shy

They're nonmetallic (or metalloids)

Without them you would die

O

Oxygen 8, the fuel of life
The liquid's magnetic (and explosive)
In sets of 3 it makes ozone
Combustion, rustin', corrosive

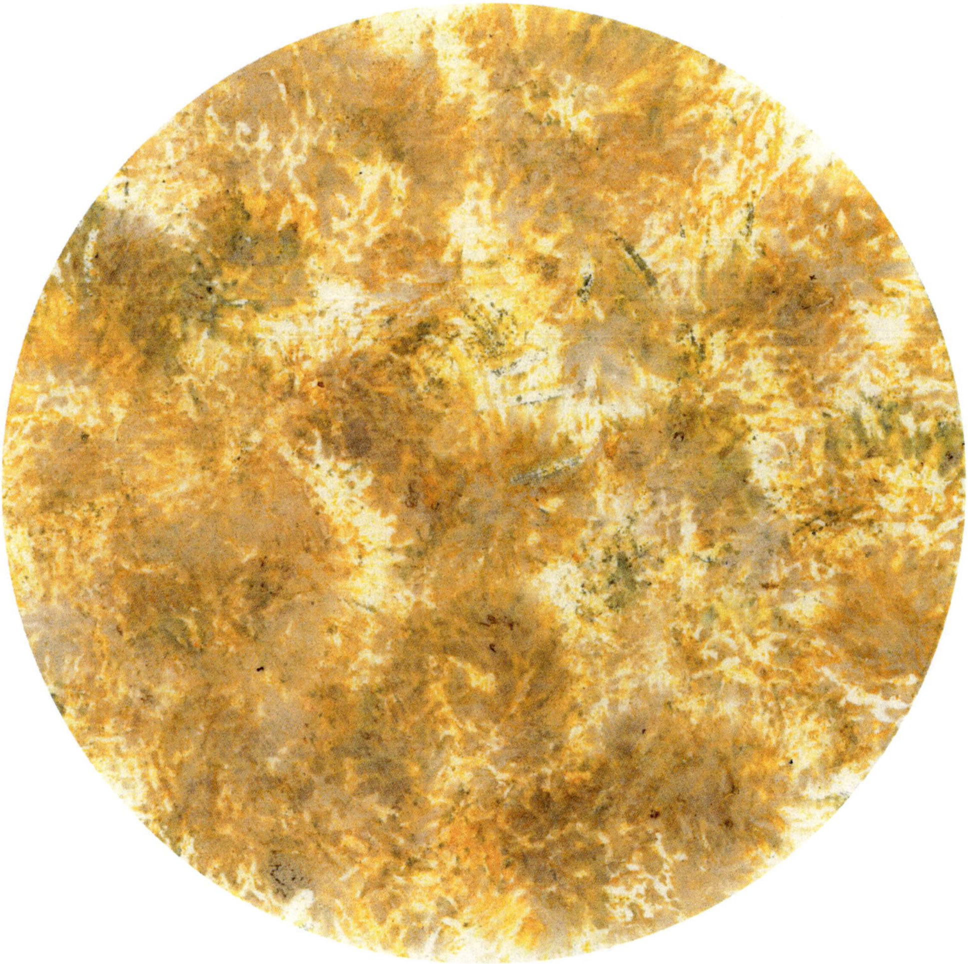

S

16 sulfur; thee who smelt it dealt it

Brimstone's yellow crystals make blue flames

Acne medication and volcanoes

Sulfuric acid→acid rain

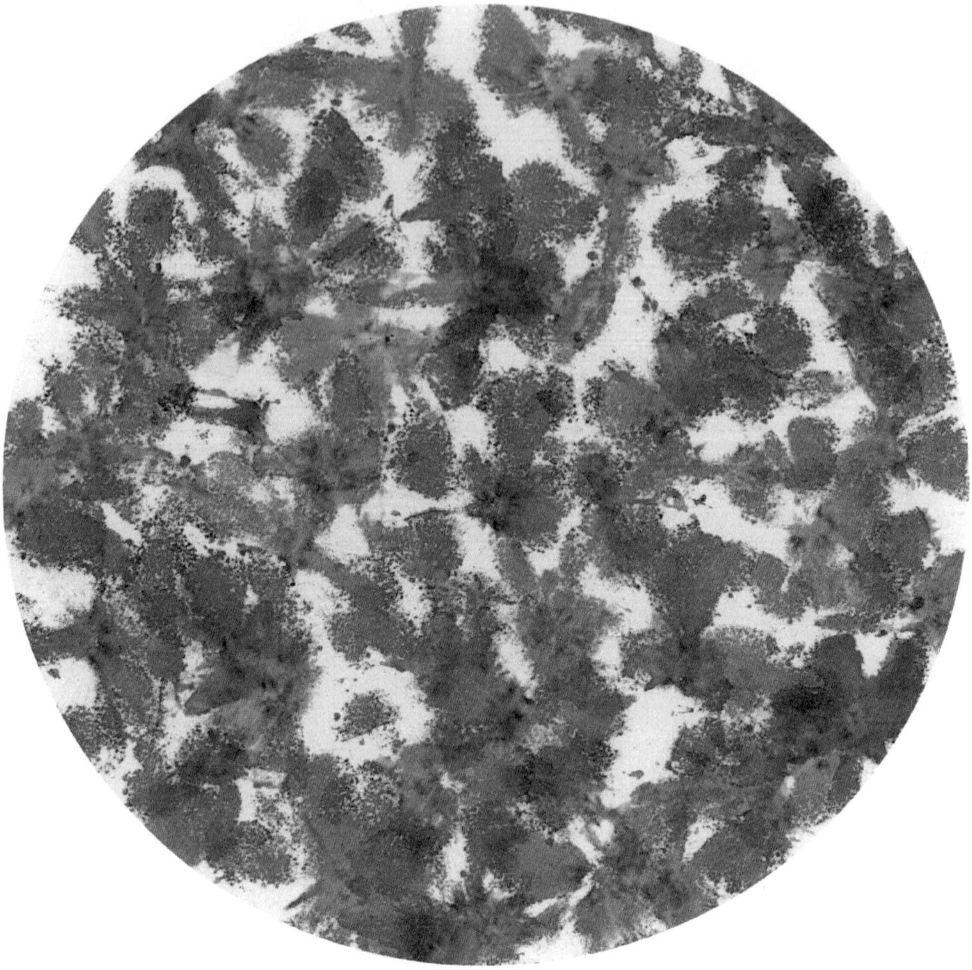

Se

Greek for "moon", selenium 34

Converts light to electricity

In Brazil nuts and laser printers

Naturally detoxifies mercury

Group 16

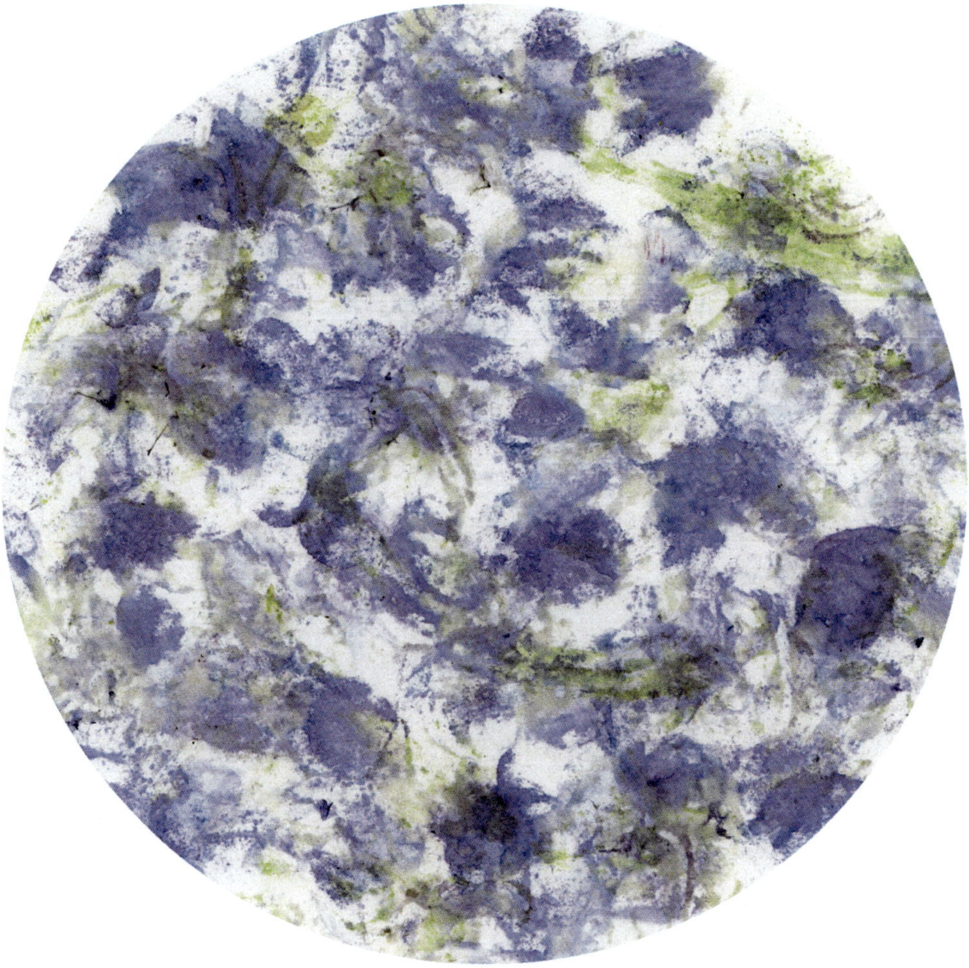

Te

52 is tellurium

Found with gold and quite uncommon

On DVD and Blu-ray

Makes your breath and BO rotten

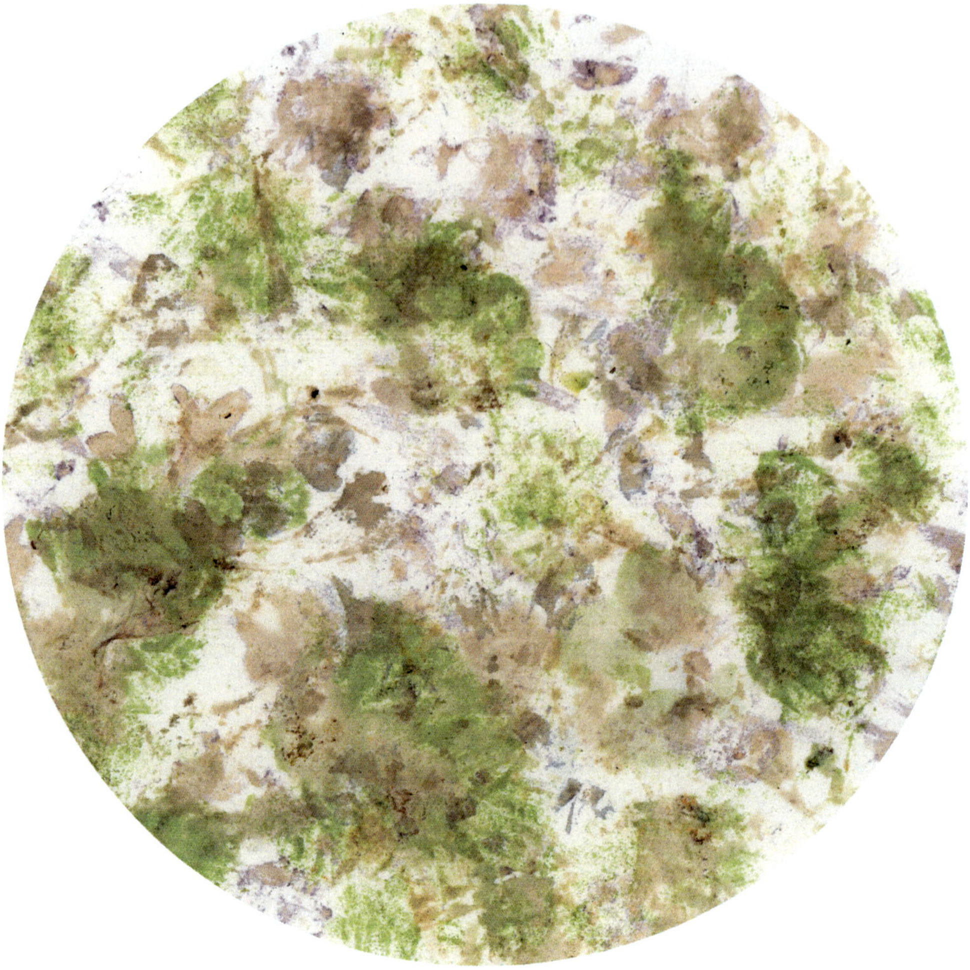

Po

Curie's first new element was polonium
One of the most toxic substances known
84 heats rockets, kills spies(?), it's in brushes
Gets excited and makes a blue glow

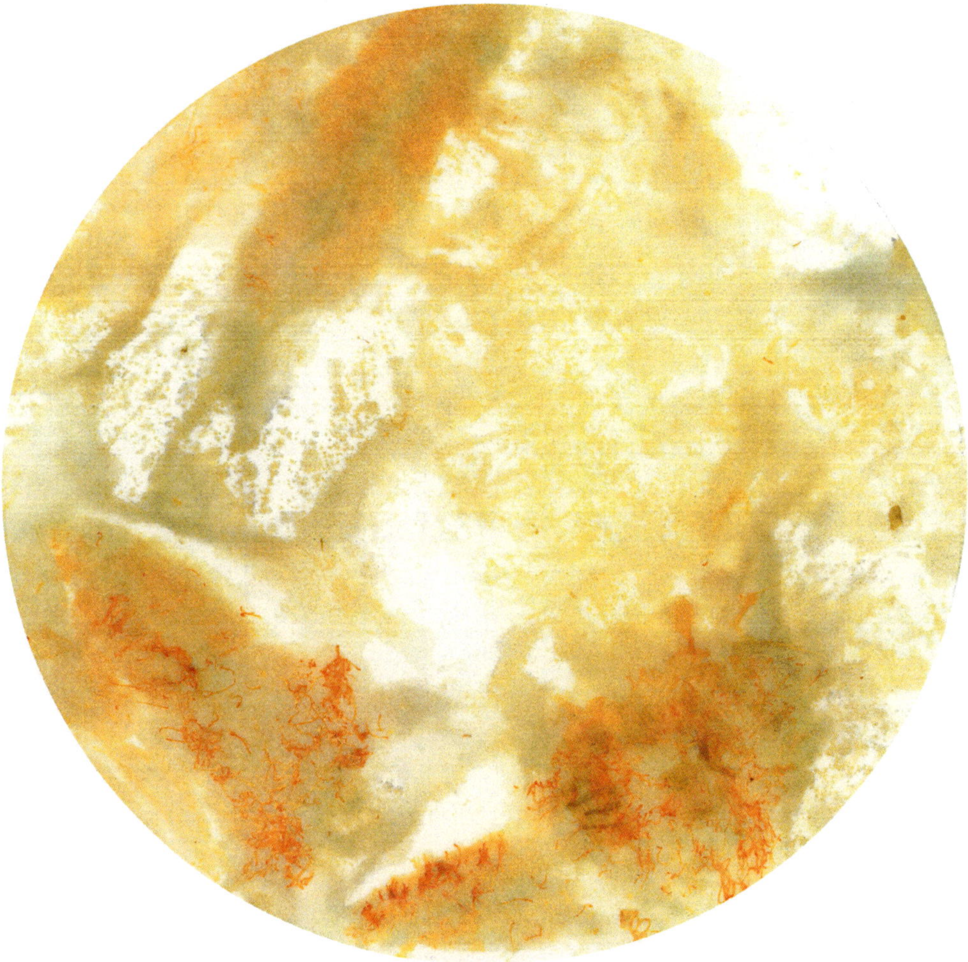

Lv

116 livermorium

Identified by its decay

A 50 millisecond half life

It cost several million to make

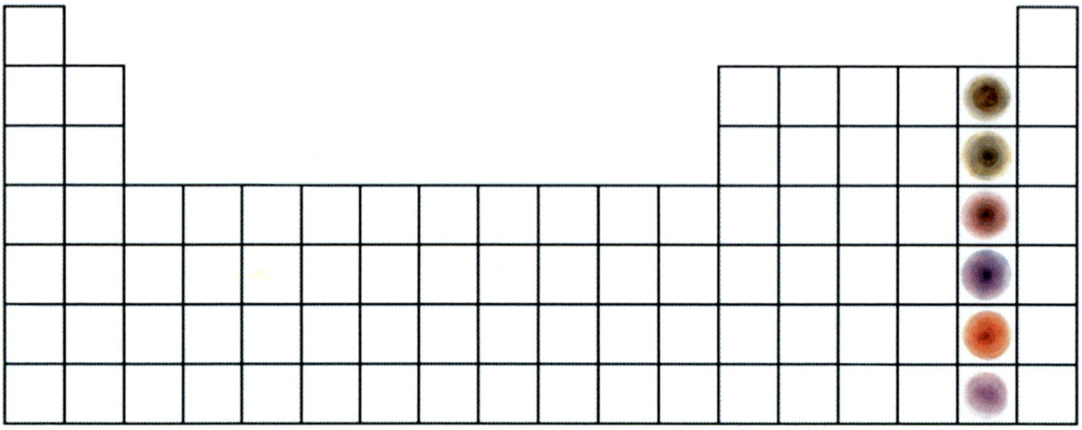

Group Seventeen

The halogens need one for an octet
Giving them extreme reactivity
Each element has a distinctive shade
And high electronegativity

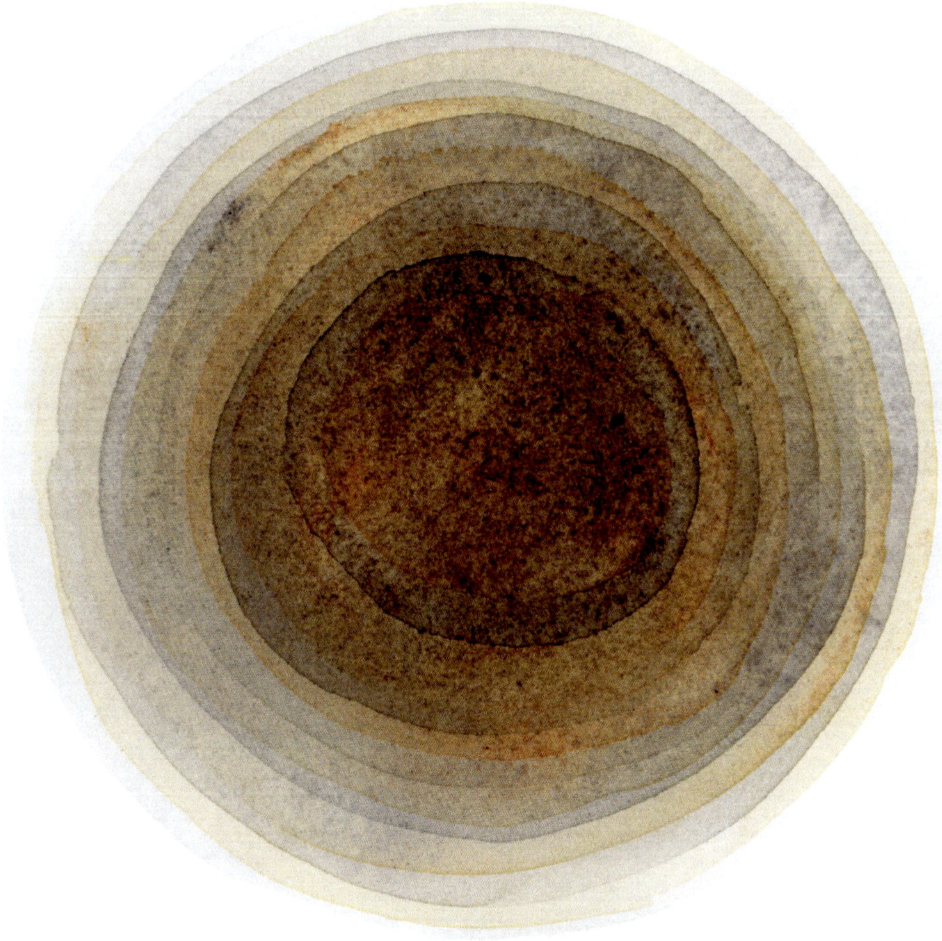

F

Fluorine 9's the most reactive
It's yellow when it's gas
In toothpaste and in Teflon
It even dissolves glass

Group 17

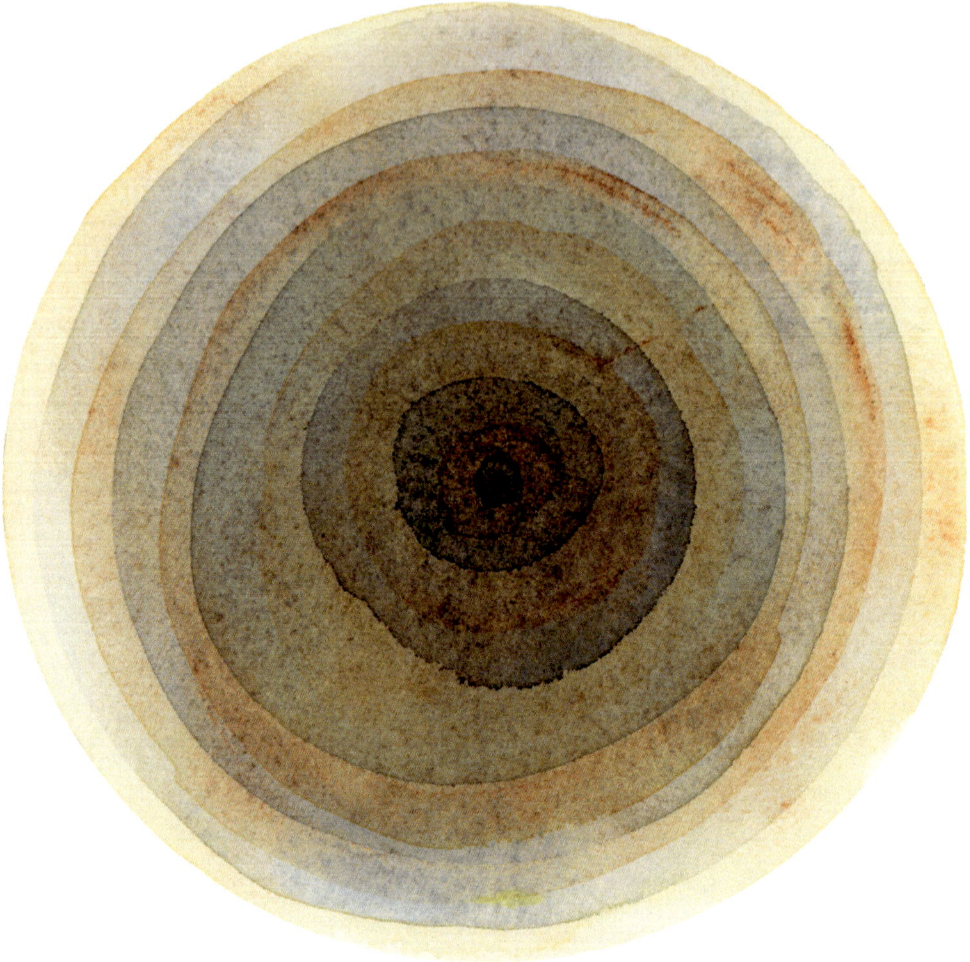

Cl

Element 17 is chlorine

Found in salt and swimming pools

CFC, PVC, DDT

War time weapon deemed too cruel

Group 17

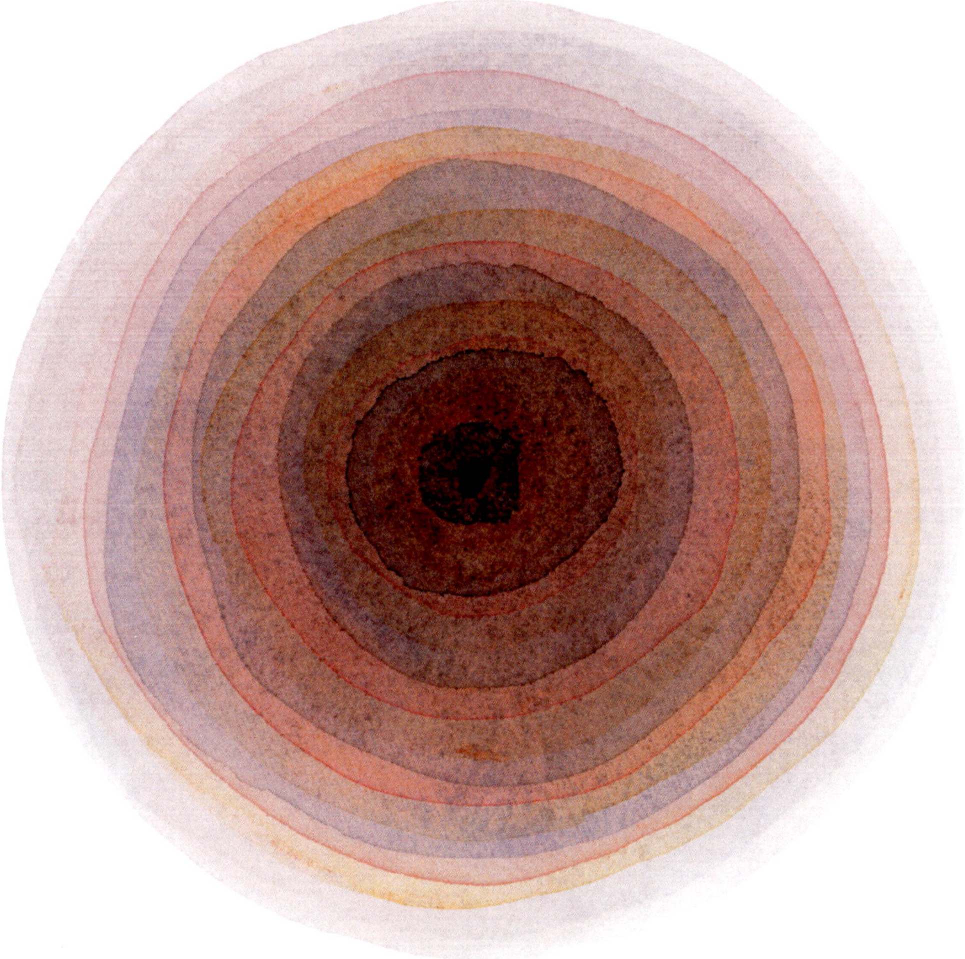

Br

Liquid element bromine 35
Evaporates in about a minute
In Mtn Dew, PJs, and hot tubs
Dyed royal Romans' purple fit

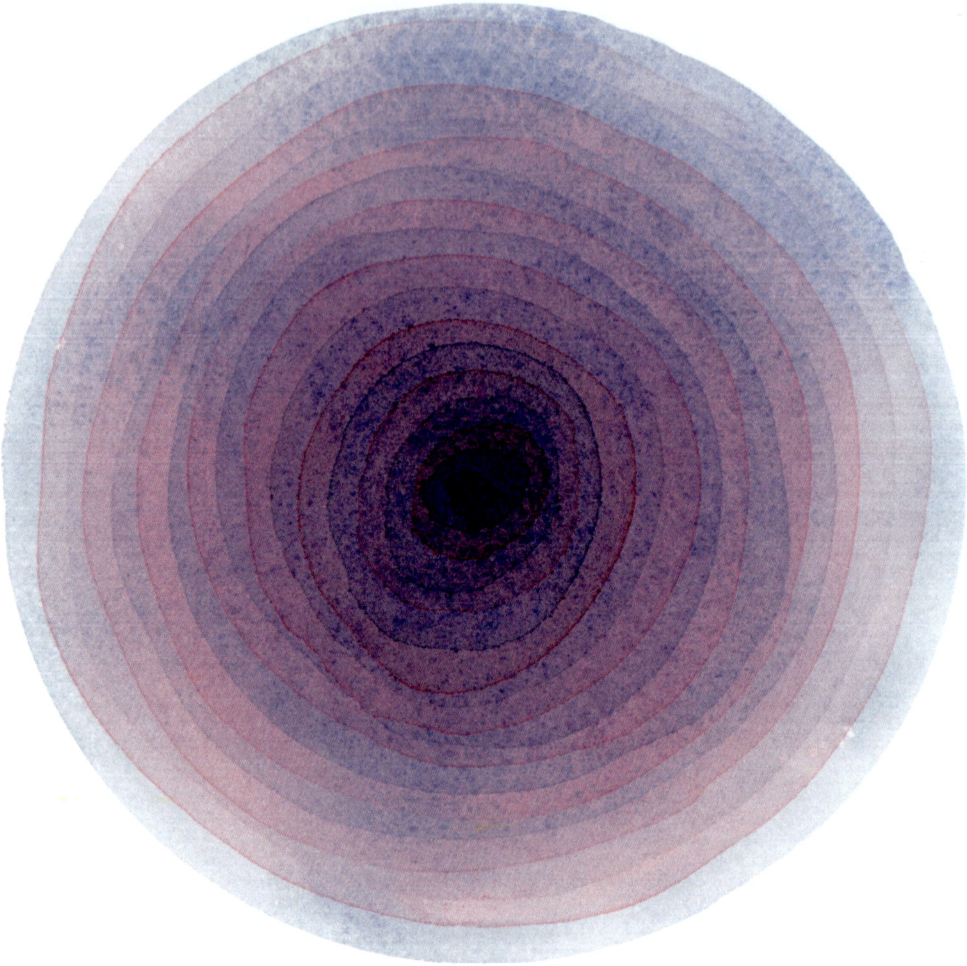

I

Iodine is element 53

Sublimation makes purple vapor

It's antiseptic for your cuts and scrapes

Prevents goiters with your salt shaker

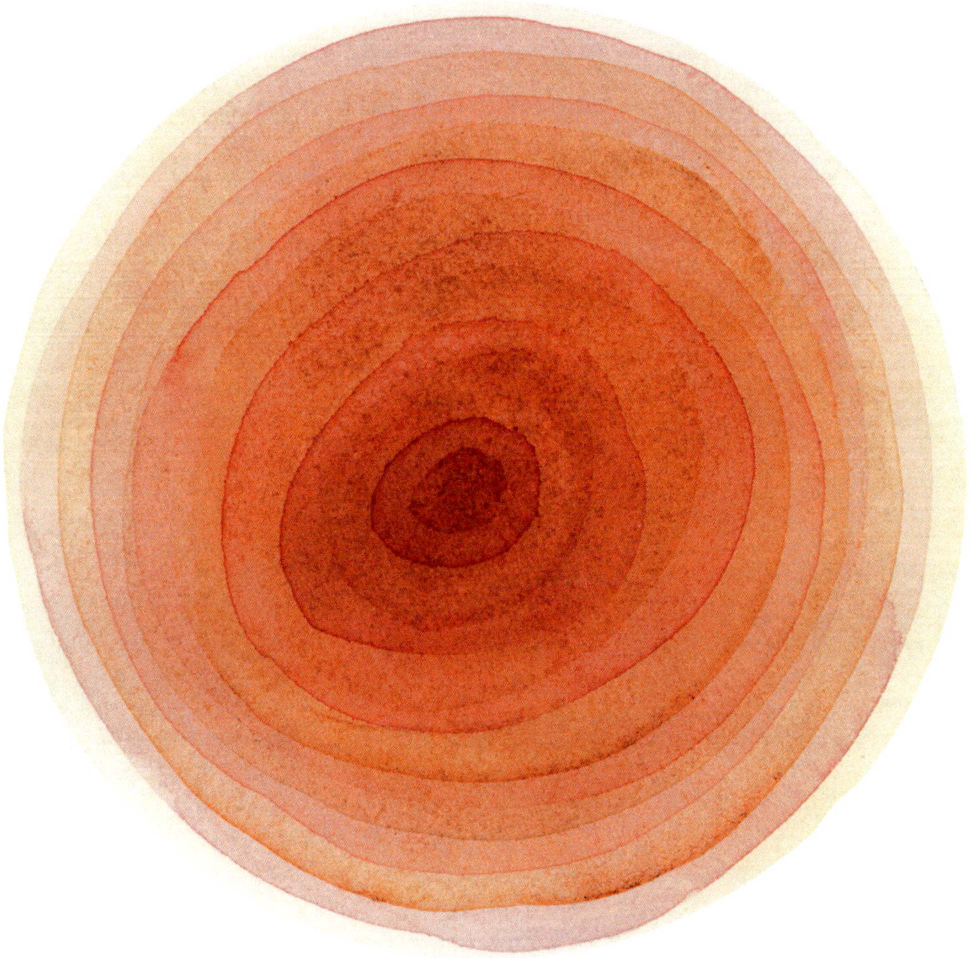

At

Astatine is element 85

Half life; 498 minutes

Potential for cancer treatment

Earth has about one ounce in it

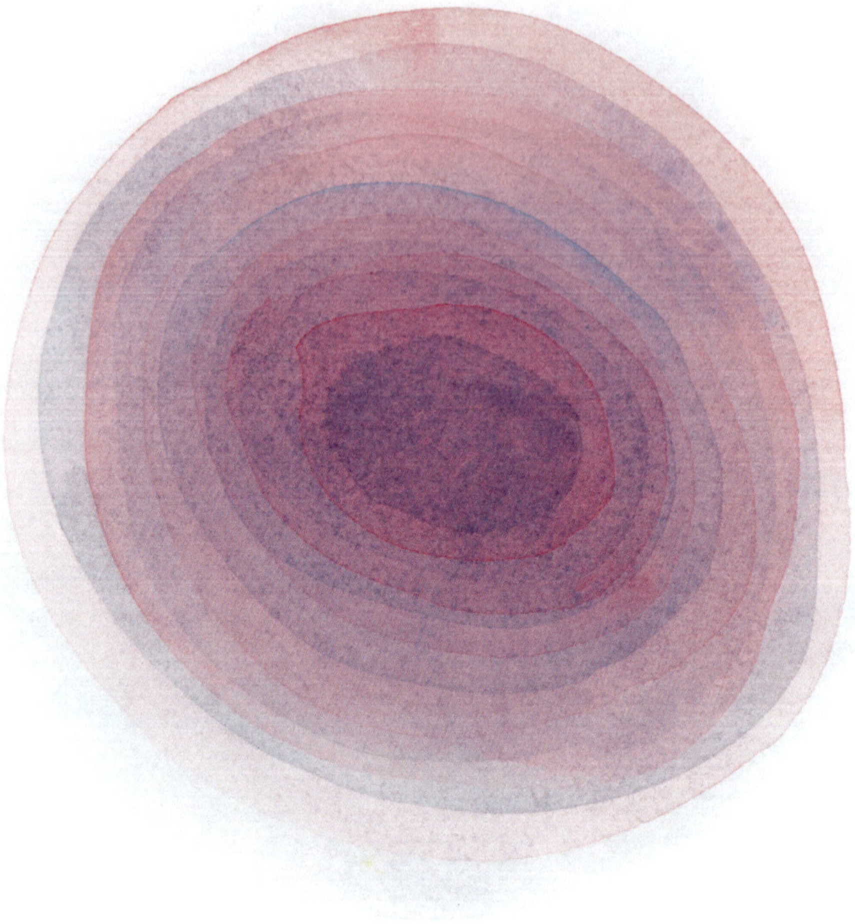

Ts

Tennessine 117

Life is short, it's not been seen

Not accepted 'til '15

Making it took lots of green

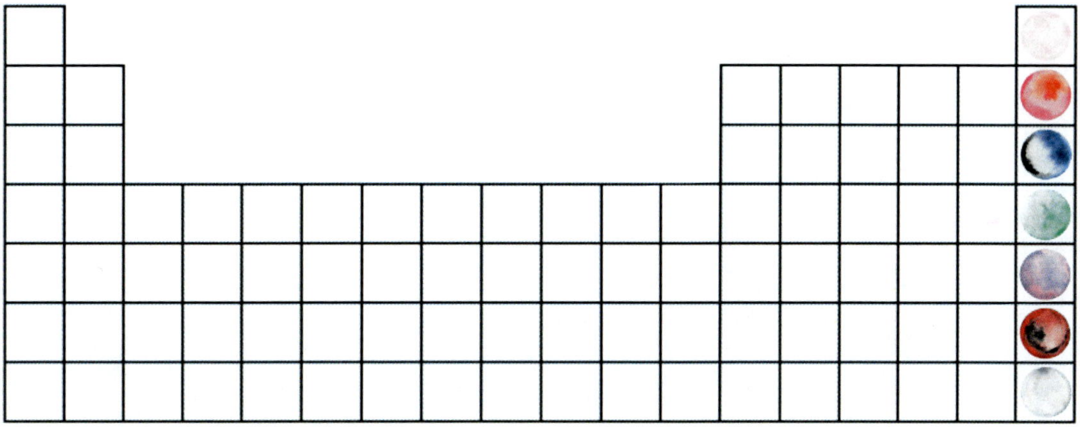

Group Eighteen

The Noble Gases are full - and extra
Too snobby to form bonds with others
(Most won’t even be radioactive)
Poor conductors but glow gorgeous colors

He

Number 2 named for Helios

Made of radioactive decay

In weather balloons and welding

Inhale and start squeaking your name

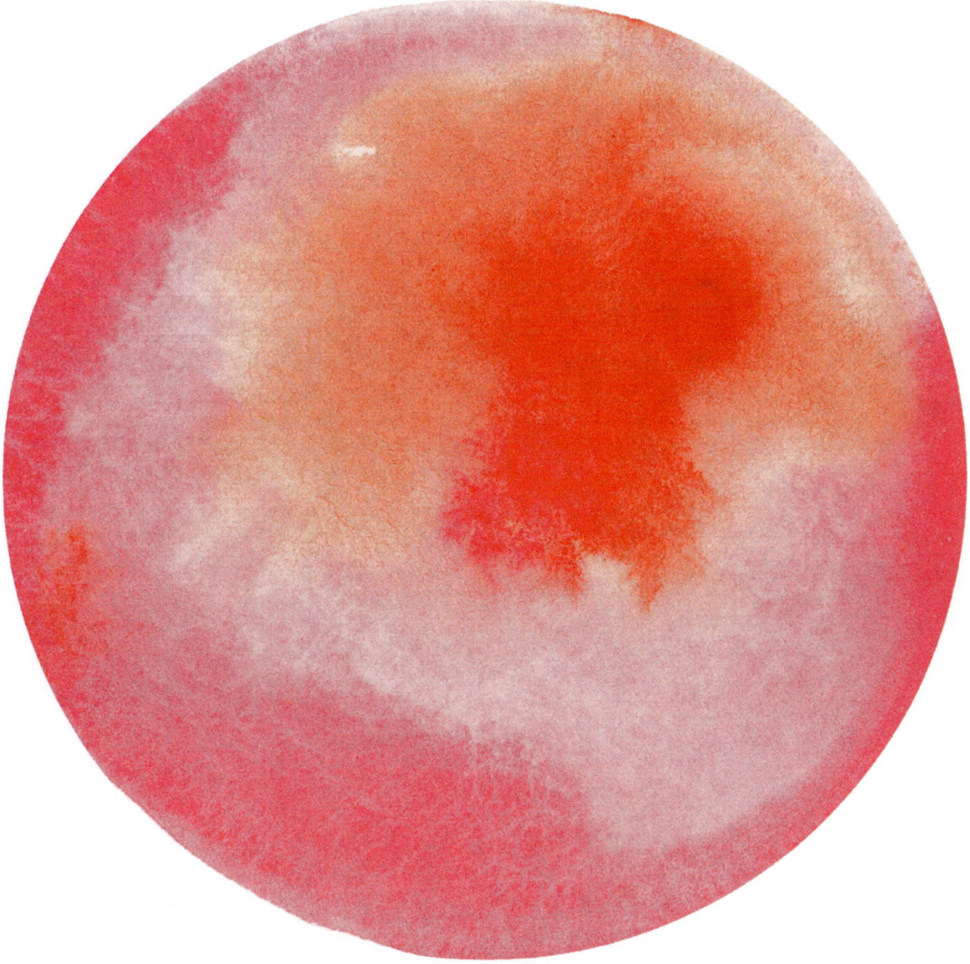

Ne

Element 10 is Neon

It only has two tricks:

Lights up the world with a reddish glow

And absolutely will not mix

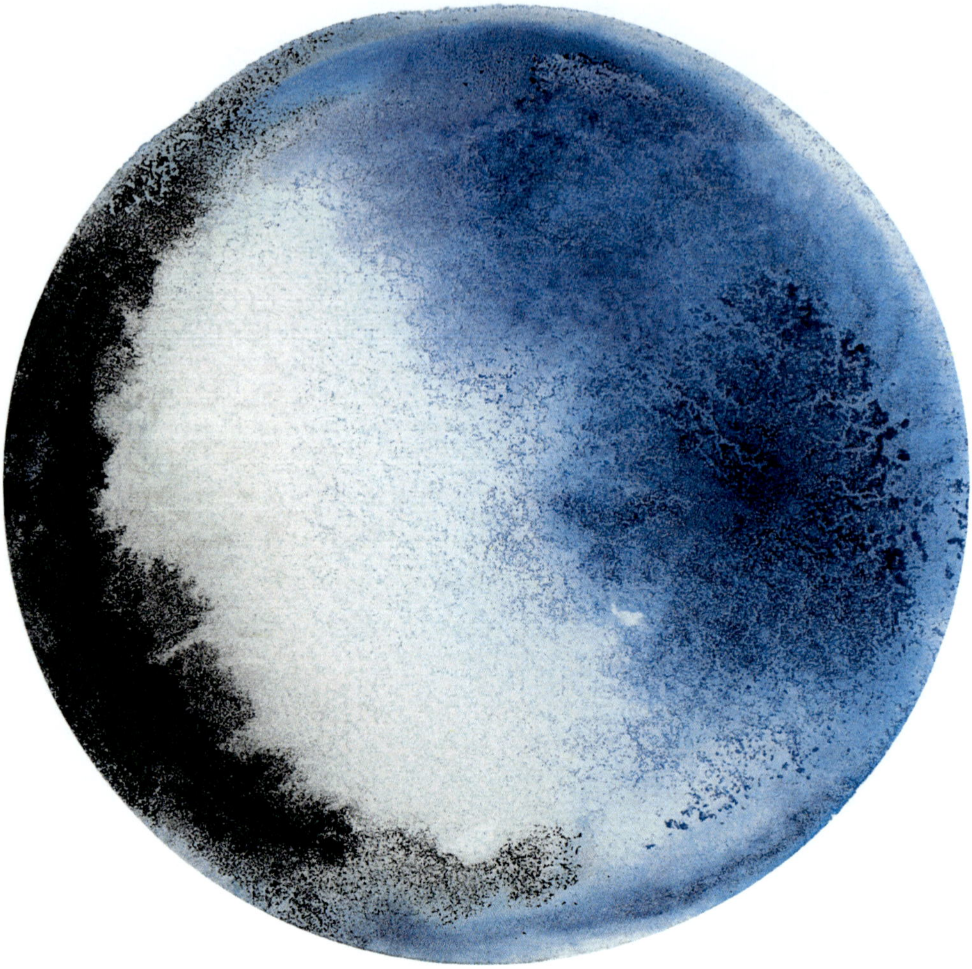

Ar

18: Argon's the 3rd most abundant
It won't react or bond - just exists
In light bulbs and LASIK and welding
It's the air that cushions your chips

Kr

Superman's world: Krypton 36

Tracks bombs, once established the meter

Inert, invisible, glows green

Chilling in cryo deep freezers

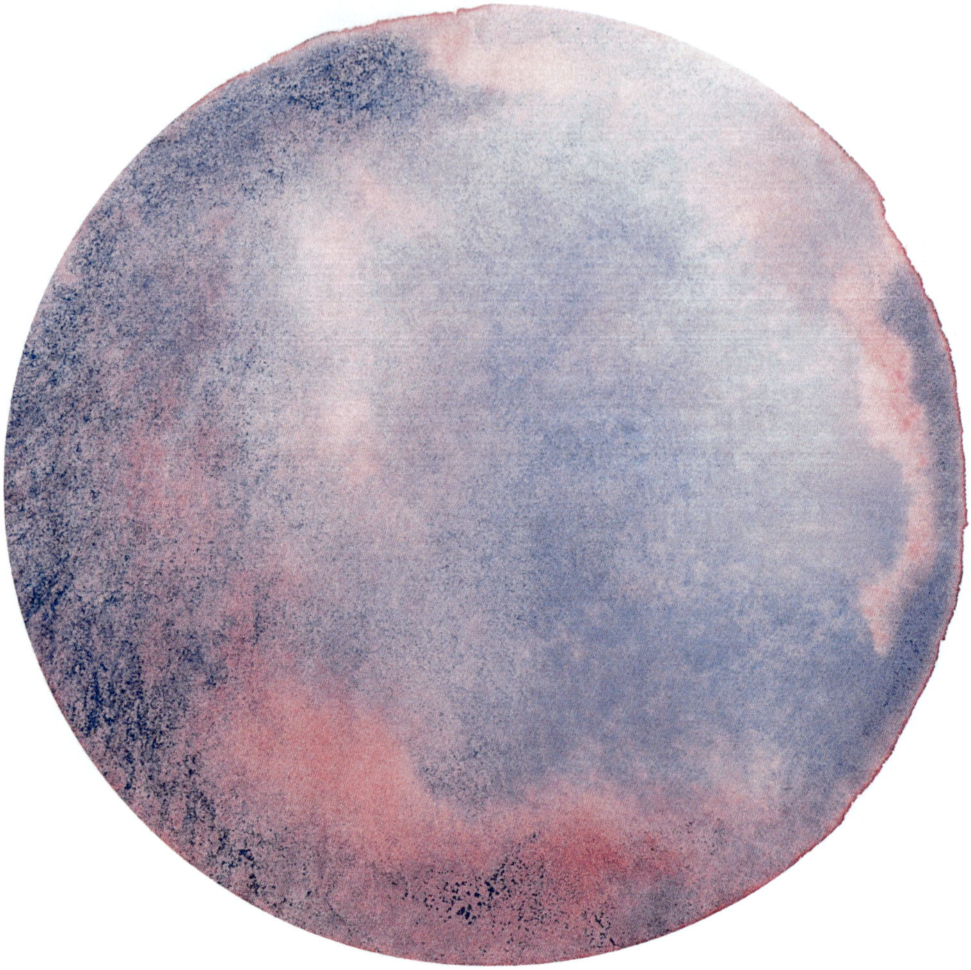

Xe

Element 54 is xenon

In bonding - it's sort of receptive

Used by IMAX and in fancy headlights

Of its group, it's the most expensive

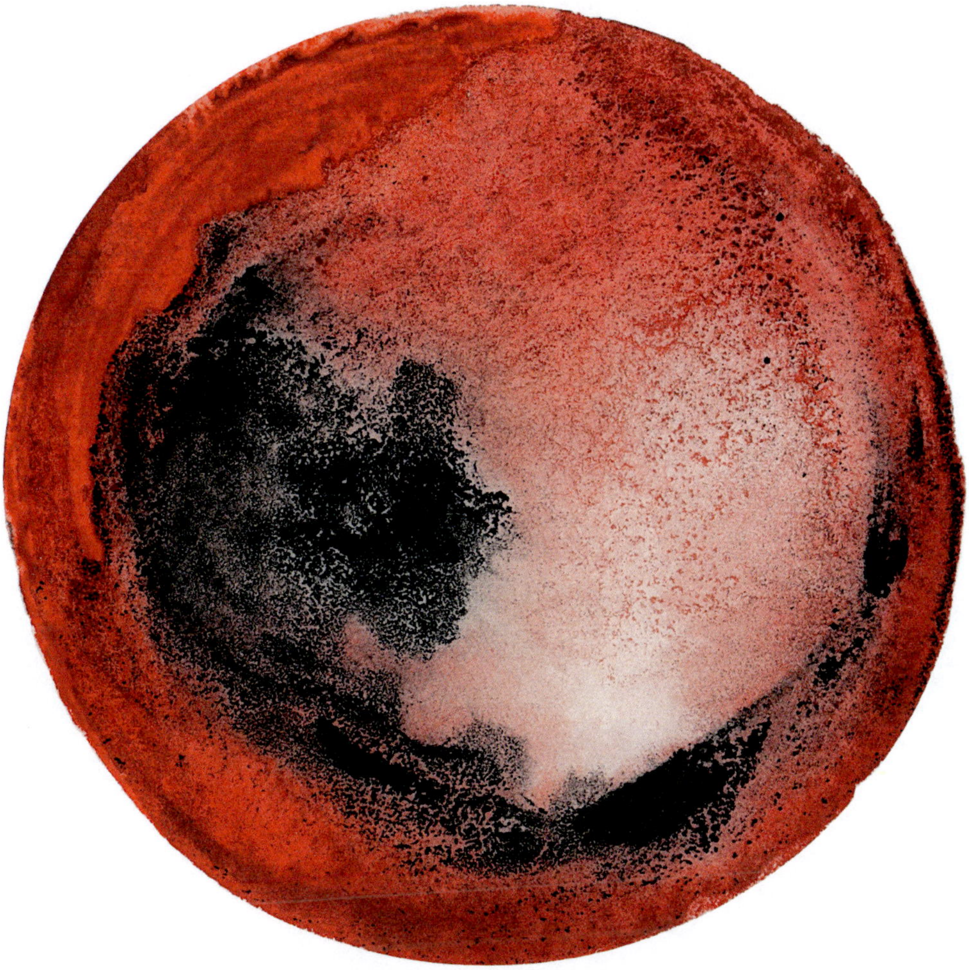

Rn

It once was a hot health fad
It's radon 86
It loves to seep in your basement
Get testing so you don't get sick

Group 18

Og

118 is oganesson

Just 5 atoms (and countless attempts)

Its atomic mass is 294

Probably solid at room temp

The Periodic Table

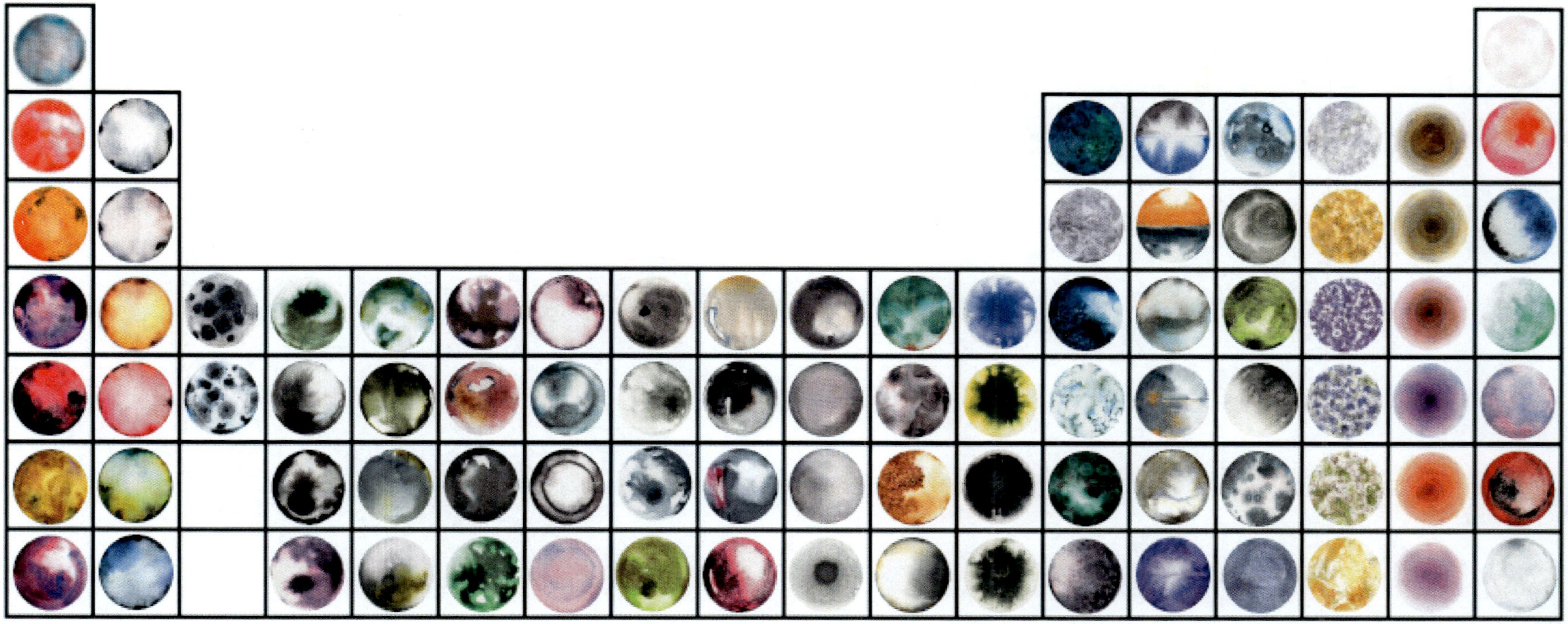

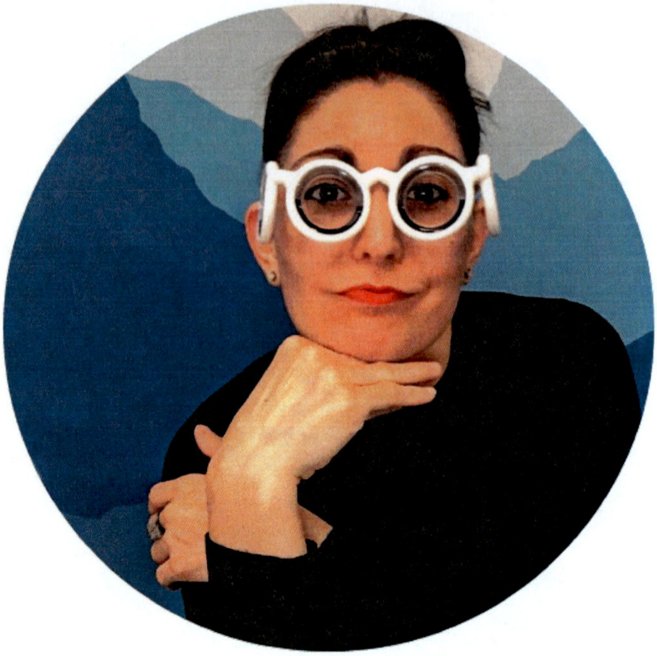

About the Author

Becky L Duncan is a fully licensed adult who writes poems about science, history, farts, and other unstable forces.

She paints with watercolors like she's running out of time and firmly believes Cinnamon Roll Sundays are both a ritual and a right.

This book is her love letter to science, chaos, and anyone who's ever been told they're "too much".

She did not love high school chemistry, lives in the Midwest, and is no longer allowed to shop in art supply stores unsupervised.

This book was powered by curiosity, kettlebells, and unreasonable amounts of sourdough.

CREDITS

Written, illustrated, and designed by Becky L Duncan

Published by Donkey Drove

First edition

Website: mzdonkey.com

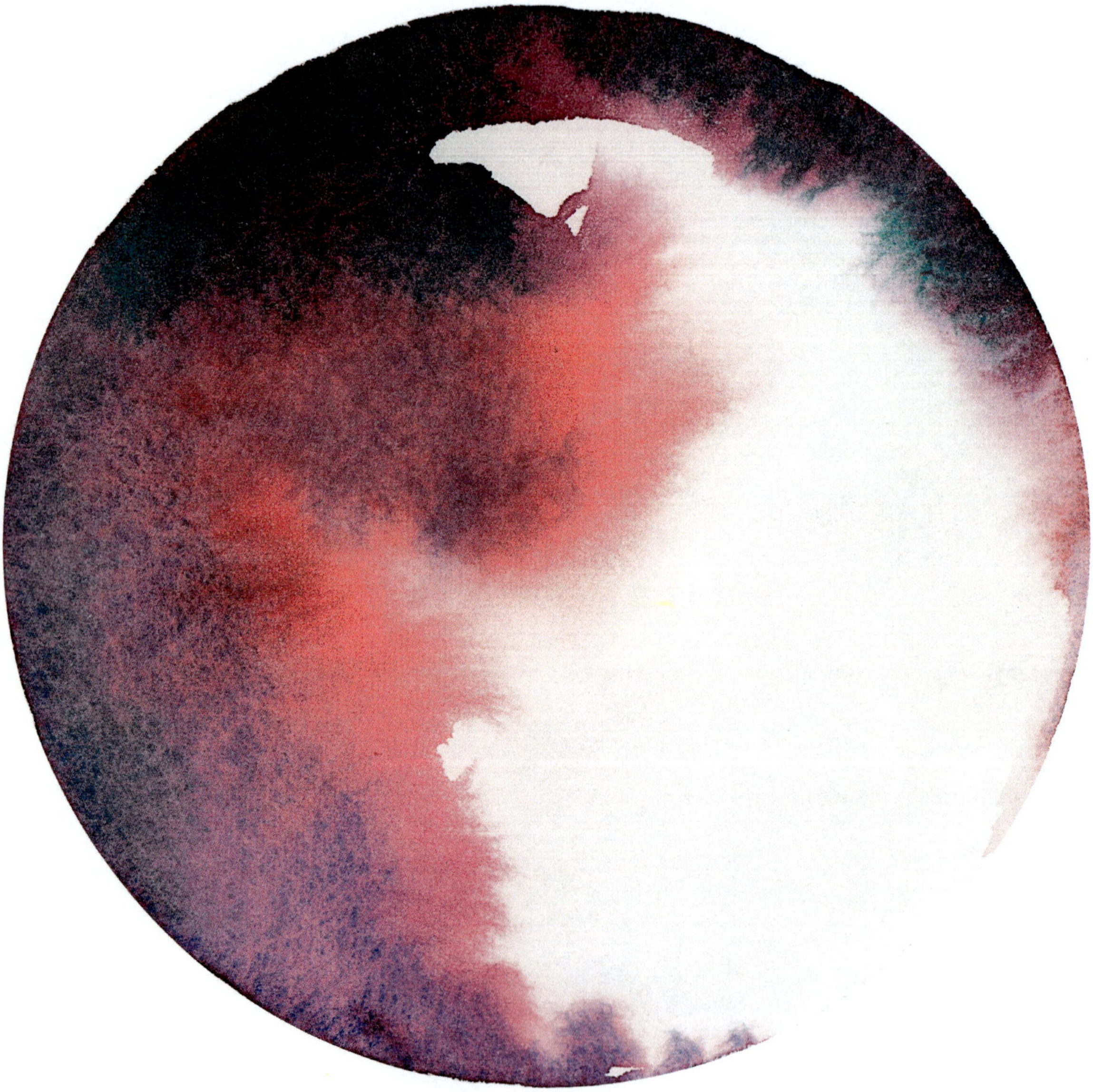

Made in United States
North Haven, CT
05 November 2026

21765023R60066